Praise for Fractal Holographic Universe

By Billy Carson

“A fascinating journey from the mathematics of fractals to the physics of the universe and the nature of consciousness. Billy Carson uses his broad-ranging vision to enrich the reader with philosophical and spiritual insights. An exhilarating read!” -**Avi Loeb**, Bestselling Author of *Extraterrestrial* and Director of Harvard's Institute for Theory and Computation

“Fractal Holographic Universe seamlessly unifies the realm of our ever-expanding consciousness with the real science of the truth that each of us is a holographic expression of the entire universe. Billy Carson has done it again.”
-**Michael B Beckwith,** Founder of Agape International Spiritual Center, Author of *Spiritual Liberation, Life Visioning, and The Answer is You.* https://agapelive.com/

“As Above So Below” takes on a whole new meaning in Billy Carson's new book “The Fractal Holographic Universe.” It offers a fresh and fascinating perspective on the universal laws that govern all aspects of life:

from the "As Above So Below" takes on a whole new meaning in Billy Carson's new book "The Fractal Holographic Universe." It offers a fresh and fascinating perspective on the universal laws that govern all aspects of life: from the micro to the macro, from the human brain to the very fabric of spacetime. You will feel inspired and awe-struck at the thought that we may have found more profound meaning to our existence and place in the universe." -**Caroline Cory**, Author and Award-Winning Filmmaker, Consciousness Science & Spirituality, *Superhuman: The Invisible Made Visible, A Tear In The Sky*, Omniummedia.com

FRACTAL HOLOGRAPHIC UNIVERSE
THE MATRIX CODE REVEALED

BILLY CARSON

$$z_{n+1} = z_n^2 + c$$

4BIDDENKNOWLEDGE
WESTON, FL

ISBN: 979-8-9871224-8-8
LCCN: 2024939563

Books may be purchased by contacting the publisher and author at:

4biddenknowledge Inc.
2645 Executive Park Dr, Suite 419
Weston, FL 33331

http://4BK.TV
4biddenknowledge.com

Info@4biddenknowledge.com

Interior formatting by Laura C. Cantu
Cover art created by Laura C. Cantu

About the Author

Billy Carson - Founder and CEO
4BiddenKnowledge

Billy Carson is the founder and CEO of 4BiddenKnowledge Inc. and a 4X Best-Selling Author of *The Compendium Of The Emerald Tablets, Woke Doesn't Mean Broke, The Epic of Humanity*, and *Fractal Holographic Universe*. Additionally, he won the 2022 Stellar Citizens Award.

Mr. Carson is also the founder and CEO of *4BiddenKnowledge TV*, a conscious streaming TV network, and the host of *Anunnaki: Ancient Secrets Revealed*, a new TV series that explores the ancient past from a fresh perspective with incredible insights into the development of an ancient global civilization.

Mr. Carson hosts the 4biddenknowledge Podcast and is the Co-Host of the Bio-Hack Your Best Life Podcast. He is also a writer and contributor to Rolling Stone and Entrepreneur magazines, and Bizjournals.com.

Mr. Carson appreciates the dedication and hard work it takes to accomplish great things. He has earned a Certificate of Science (with an emphasis on Neuroscience and Artificial Intelligence) at MIT and a certificate in Ancient Civilization from Harvard University. Among his most notable achievements, Billy is the CEO of First-Class Space Agency, based in Fort Lauderdale, FL. His space agency is involved in researching and developing alternative propulsion systems and zero-point energy devices

Expert host on the following networks:

4biddenknowledge TV, History Channel, Travel Channel, Discovery Channel, Discovery Plus, Science Channel, ABC, CBS, Gaia

For more information visit IMDB.com

Featured in:

Bloomberg, Businessweek, Rollingstone, TIME, Forbes, USA Today, Entrepreneur

Documentaries and TV Shows:

- 2024
  Anunnaki: Ancient Secrets Revealed

- 2023
  After Contact
  1st Annual 4didden Conscious Awards

- 2022
  Black Knight Satellite

- 2021
  4biddenknowledge Podcast
  Bio Hack Your Best Life Podcast

- 2019
  UFO's: The Lost Evidence (TV Series)
  World's Biggest Mysteries (TV Series)

- 2017
  What If (Documentary)
  DocUFObia (Documentary)
  Ancient Civilizations (TV Series)
  Beyond Belief with George Noory (TV Series)

Life Beyond Our Existence (Documentary)
Buzzsaw with Sean Stone (TV Series)

- 2016

  The Anunnaki Series (TV Series documentary short)
  Deep Space (TV Series documentary)
  Baltic Sea Anomaly: The Unsolved Mystery (documentary)

- 2015

  UFAH Favorites (Video short)

- 2012

  Countdown to Apocalypse (TV Mini-Series documentary)

Dedication

Billy and Elisabeth Carson

To my wife, Elisabeth Carson.

Your strength and perseverance inspire me every day to keep becoming a better version of myself. Thank you for being my constant source of support and motivation. I love you dearly.

Also by Billy Carson

Check Out Other Works by 4BiddenKnowledge

Download the 4biddenknowledge TV App

Scan the QR Codes to Enjoy These Creations by Billy Carson, CEO and Creator of 4Bidden Knowledge

Contents

Foreword

Robert Edward Grant - Award Winning TV Host, Code X streaming on Amazon Prime Video and Gaia TV - 2x Best Selling Author - Polymath, Field Researcher - Entrepreneur and Thought Leader

For many thousands of years, mankind cast its gaze into the night skies and asked its most existential questions: Who are we? Why are we here? Are we alone? Where are we going? And what if at all, is the purpose to life? In an attempt to make sense of what can at times appear to be a somewhat senseless human story, a rich mosaic of cave paintings, lore, mathematics, time, measurements, mythology, civilizations, religious and spiritual texts, history, science, and psychology have emerged as artifacts that inform us along our collective journey toward answering such vexing questions.

In the 4th Century BC, the Polymath and Philosopher Plato posited a new theoretical thought model of perceiving the Universe in his now very well known "Allegory of the Cave" within his book Republic. While also harboring political overtones of his (Plato's) day, this novel approach to perceiving the Universe notionally presented a reflected and absorbed (shadow) reality (including an inverted U-inverse), a worldview for what we call reality that was perhaps only a shadow facsimile of another more "original" projection. That the experience we are living is merely a holographic projection of some other more fundamental and primary experience that actually has been interconceived of a Universal "Mind." The Materialists that followed Plato perceived this distinction as a more metaphorical and, therefore, more 'philosophical' theory of everything rather than a literal explanation of the Human experience altogether that could be proven by reductionistic and empirical experimentation.

The 20th Century brought major scientific advances to our understanding of "classical" (the Standard Model of) Physics. For a few decades, the Material Sciences and other related technological advances seemed almost unstoppable, raising new existential questions about Humanity's ability to survive its own creations in a Nuclear Age. At approximately the same time, the advent of a new and emerging model for Physics presented a novel concept of Mentalism that seemed to take Humanity's leading thoughts to a more literal conception of a Holographic Projected Experience (arguably more aligned to Plato's original Cave allegory). The advent of Quantum Physics also introduced a new variable to its objective 'material' experience: a conscious observer. Furthermore, the "Double Slit" (including the Quantum Erasure//Delayed Choice variants) experiments raised problems within the Classical model of Physics, introducing what Albert Einstein PhD would later refer to collectively as "spooky action at a distance" in describing entanglement as well as the role of a conscious observer to impact the outcome of experimentation generally. Meanwhile, his colleague and mentor, physicist Max Planck PhD, famously concluded, "There is no matter as such...We must assume behind this force the existence of a conscious and intelligent mind. This mind is the matrix of all matter." By the mid-twentieth Century, it certainly seemed that the Materialists' view of the Universe as an objective materialism model of reality was soon to be under more scrutiny and could even be replaced with a new and almost unthinkable model for our Universe: a "subjective mentalism" based model of reality.

Twentieth Century contemporary and noted leader in Analytical Psychology, Carl Jung PhD presented an amalgamation of mythological archetypes and the division of conscious, subconscious/unconscious approaches to mankind's quest to understand his reality and even the effect of his observation upon his experience. He even worked closely with Albert Einstein PhD and Wolfgang Pauli PhD in an attempt to interpret mankind's "reality," which appeared to be more and more related to its perception and bias than had been previously objectively considered. He also introduced concepts of reflection and shadow to interpolate his paradigm of a Conscious Universe.

More recently, the "Simulation Hypothesis," introduced by Oxford University Philosophy Professor Nick Bostrom, PhD, has emerged as a viable explanation of this interactivity of the Conscious Observer and the Observer's actual experience. Once seen as an impossible theoretical framework as a "Theory of Everything," more and more, both within (and without) the academic community, are siding with this controversial yet compelling theory that mankind is experiencing its own fractal holographic simulation and experience.

In 2022, the Nobel Prize was awarded for proving entanglement in experimentation in Stockholm, Sweden, which struck a serious blow to the Materialist's view of an objective and material reality among academic circles and the physics community at large. Indeed, the headline article in Scientific American declared: "Local Realism is False, and this Year's Nobel Prize Winners Proved it." While many questions remain, this was perceived by many as yet another convergence toward a Simulation Hypothesis that is entirely more subjective (at all scales of observation) than previously considered.

Fractal Holographic Universe takes the reader on a journey and dives deeply into this analytical framework of the Human condition and experience. It takes on deeply held taboos and challenges the reader to comprehend a new thought model and perhaps even unlearn some of the dogma we may have unknowingly accumulated and subscribed to along our life's journey.

In his customarily pragmatic and common sense approach, Billy Carson taps his broad understanding across many academic disciplines, including physics, mathematics, history, ancient mysteries, and esoteric knowledge, to take on the oldest questions of our Human experience: what it is and perhaps even address some of the more fundamental "why" questions as well. Billy seamlessly draws upon the expertise of many luminaries within the fields of mathematics, physics, and philosophy to augment his deep knowledge, making it a credible and enjoyable read for virtually any student of life asking the existential questions of our existence.

Mankind may never fully answer our most vexing questions of who we are, why we are here, and even our ultimate reason d'etre, and even if we do, it is possible that more new questions will only come as a result of what we find. As it has been said so many times (and in so many different ways): the more we learn, seemingly, the less we actually know; the Universe and our experience within it may not only be stranger than we imagined, but rather even stranger than CAN be imagined. Just like life itself within the Fractal Holographic Universe, perhaps what matters most is the journey of self-realization undertaken above, arriving at the destination itself?

Robert Edward Grant

http://robertedwardgrant.com

Introduction

After decades of extensive research on the nature of reality, I have concluded that we exist within a fractal holographic universe, a concept that reaffirms our reality rather than negating it. This modern concept reflects the wisdom of ancient civilizations by providing a scientific way to explain how things are created in our world and the basic code that governs it. The Mandelbrot set is an intriguing pattern that exists within a vast two-dimensional area called the complex plane. The Mandelbrot set emerges when a specific mathematical operation is repetitively applied to numbers within this plane, causing those outside the set to diverge toward infinity, while those within exhibit a kind of controlled chaos, dancing and drifting near its boundaries. This boundary region is characterized by intricate, carefully orchestrated movements indicative of an onset of instability.

I'm in a photo at the Field Museum in Chicago, standing in front of the Mandelbrot set, a complex and beautiful mathematical pattern. The set, named after Benoît B. Mandelbrot of the IBM Thomas J. Watson Research Center in Yorktown Heights, New York, is a cornerstone of fractal geometry. Mandelbrot's research on geometric forms paved the way for studying shapes with fractional dimensions. The Mandelbrot set's boundary is more than just a fractal.

Field Museum in Chicago – Billy Carson

The intricate patterns of the Mandelbrot set deeply intertwine with the Adinkra codes, discovered by Professor Sylvester James Gates Jr. The Adinkra codes and patterns reflect the principle of "as above, so below," implying a mirroring between the macrocosm and the microcosm, between the universe and the minutiae of its individual components.

In essence, we are living in a reality that can be likened to a created light matrix. This perspective doesn't just offer a glimpse into the complex beauty of the universe; it provides a profound understanding of the underlying structure of our very existence.

The concept of the third dimension posits that it is a holographic projection emanating from a two-dimensional plane situated at the periphery of our universe. In this framework, our spirit is perceived as being entangled within this third-dimensional holographic realm, composed fundamentally of atoms. The human brain functions like a crystal receiver, turning electromagnetic signals into holograms.

Quantum mechanics introduces the idea that an atom exists in a state of probability, distributed across space, until it is observed by a conscious observer. This act of observation or measurement precipitates the manifestation of the entire universe. This theory suggests that conscious beings are fundamentally interwoven with the existence of reality itself. Without conscious observers, the universe would merely be an expanse of unmanifested potentialities, with no definitive events occurring.

Delving deeper into the nature of what is perceived as matter, it becomes increasingly apparent that, at a fundamental level, things are far less solid than they seem. What we perceive as the real world is actually an illusion. We know that atoms are 99.999% empty space, which suggests there is no true solidity.

Consequently, distance and separation are illusions as well. Everything in the third dimension is energetically connected, and this connection is continuously expanding, originating from one Source—the beginning of everything in this realm.

It seems we have developed illusions of distance, location, separation, individuality, and even the arrow of time to navigate the lower density of the third dimension. However, from higher dimensions, the past, present, and future can be seen and experienced all at once.

Thus, the third dimension is an illusion based on electromagnetic frequencies. The double-slit experiment, discussed later in this book, demonstrates that everything exists as a wave of potentials until a conscious observer interacts with these waves and collapses them into digital bits of information. This experiment proved that everything in this dimension exists as waves of light—light that we can't perceive because we can only see 1% of the light spectrum. Yet, it is light nonetheless. When this light slows down to a specific frequency, it gives the illusion of solidity.

For example, when I touch a wall, my hand and the wall seem solid, so it feels like I can't penetrate it. But in reality, I am not touching the wall. The electromagnetic fields created by the electrons orbiting the atoms in my hand are repelling the electromagnetic fields of the electrons orbiting the atoms in the wall. Thus, we never actually touch anything; it is the repulsion between electromagnetic fields that we experience.

Image illustrating the concept of reality as an illusion, showing a person reaching out and their hand appearing to penetrate a wall, with waves of light and energy surrounding the interaction. The background combines various landscapes and abstract representations of different dimensions, emphasizing the holographic nature of reality. 4biddenknowledge Inc

If I could match the subatomic frequency of the vibration of the atoms in the wall with my body, I could move my hand through the wall and even walk through it. On a grand scale, we must understand that this dimension could be a holographic projection that we navigate through consciousness.

The human brain processes millions of signals every minute, organizing these into holograms, which we then project externally, perceiving them as our reality.

Light Matrix – 4biddenknowledge Inc

In this context, a hologram serves as more than just a visual phenomenon; it represents a metaphorical bridge. It is the means by which we reduce "N dimensions" of information into "N minus 1 dimensions." This concept helps reconcile the paradoxes we encounter and facilitates leaps from one concept to another. While we may not be creators of reality in the absolute sense, we certainly craft our own subjective realities or "reality tunnels," many of which we are not even consciously aware of.

Every element of matter, at its core, is energy condensed to a slower vibration. This perspective aligns with the notion that all existence is part of a unified consciousness, experiencing itself subjectively. Echoing the words of the late Bill Hicks, "Ultimately, there is no such thing as death; life is only a dream, and we are the imagination of ourselves." This viewpoint invites a deeper contemplation of life, death, and the very essence of reality, suggesting an interconnectedness and oneness at the heart of all existence.

Chapter 1

From Microcosm to Macrocosm

The Concept of a Fractal

In the realm of mathematics and geometry, a fractal is a term that represents a never-ending pattern, one that repeats itself at different scales. This concept is known for its inherent complexity and how patterns emerge and recur at smaller scales. The mathematician Benoît B. Mandelbrot coined the term "fractal" in 1975. It comes from the Latin word fractus, meaning broken or fractured.

At the heart of fractals lies the concept of self-similarity, where a structure is made of smaller copies of the whole. In essence, if you zoom into a fractal, you will see the same shape or pattern, no matter how deep you go. This property is not limited to exact replication; sometimes, the self-similarity is approximate, but the overarching theme is that the larger structure can be subdivided into smaller parts that resemble the whole in some capacity.

Fractal design based on the Mandelbrot Set – 4biddenknowledge Inc

One of the most famous and visually striking examples of a fractal is the Mandelbrot set. This particular set of numbers creates a pattern that looks like a collection of blobs when visualized, but upon zooming into any of its blobs, more and more intricate details emerge, revealing patterns that closely resemble the larger image. The beauty of the Mandelbrot set is that it's a mathematical structure derived from a simple equation, yet it has infinite complexity.

$$z_{n+1} = z_n^2 + c$$

Mandelbrot Set Equation

Nature, as it turns out, is replete with fractal patterns. Fractals have fascinated scientists and artists because they appear in many natural phenomena. The branching of trees, the formation of river networks, the intricate designs of snowflakes, and even the structure of our lungs with their bronchial branches all exhibit fractal properties. Coastlines, too, are fractal in nature. If you measure a coastline using segments of one hundred miles each and then measure it again using segments of one mile each, the total measured length will be significantly different. This is because smaller segments capture more of the coastline's detailed curves and indentations—the nooks and crannies—that larger segments overlook. This paradox, known as the coastline paradox, was one of Mandelbrot's inspirations for delving into fractals.

The importance of fractals goes beyond their ubiquitous presence in nature or their aesthetic appeal. They have practical applications in various fields of science and engineering. In physics, fractals relate to the study of chaos and complex systems. Fractals have been used in computer graphics to create complex designs and landscapes, which are vital for making movies and video games. In medicine, researchers study the fractal geometry of tissues to gain insights into various pathological conditions.

The profound implication of fractals is how simplicity can give rise to complexity. The rules that generate fractals are typically straightforward. Yet, by repeatedly applying these simple rules, you can achieve structures of breathtaking intricacy. This property stands as a testament to the principle that complexity in nature doesn't necessarily arise from complex rules. Often, simple iterative processes can give rise to the vast diversity and complexity we observe around us.

From a philosophical standpoint, fractals challenge our conventional understanding of scale and detail. In traditional Euclidean geometry, an object has a finite number of details. For instance, you'll only see straight lines if you keep zooming into a triangle. Fractals, however, introduce a new paradigm where an object can possess infinite detail. This shift has led to a deeper contemplation of the nature of space, reality, and the patterns inherent in our universe.

Moreover, the concept of fractals bridges the gap between order and disorder. While they may appear chaotic and random at first glance, they are the product of deterministic processes. This juxtaposition of chaos and order prompts us to think about the many systems in nature that, though seemingly erratic, are governed by deterministic laws that we may not yet fully understand.

Conclusion

The concept of a fractal offers a mesmerizing window into the intricate dance of simplicity and complexity. Its patterns, whether found in a mathematical set, the veins of a leaf, or the vast structures of galaxies, challenge and expand our understanding of the universe. Fractals remind us that complexity can arise from simplicity, that nature is infinitely detailed, and that there is a beautiful order even in what seems most chaotic. As we continue to delve deeper into the mysteries of our universe, the language of fractals will undoubtedly play a pivotal role in shaping our understanding.

Fractals in nature: 4biddenknowledge Inc

The Holographic Principle in Physics

In the realm of theoretical physics, few ideas have captured the imagination as vividly as the holographic principle. At its heart, the principle suggests that information encoded on a two-dimensional boundary can describe the entire universe, much like a hologram. This notion challenges our traditional perception of reality. It indicates that the three-dimensional world we observe and inhabit may just be a projection of information stored on a distant two-dimensional surface.

The roots of the holographic principle are deeply intertwined with the study of black holes. In the 1970s, Stephen Hawking and Jacob Bekenstein made groundbreaking discoveries about the entropy of black holes. They found that a black hole's entropy, or measure of disorder and information, was proportional not to its volume but to the area of its event horizon, the boundary beyond which nothing can escape. This was counterintuitive, as in everyday systems, the amount of information or entropy they can contain scales with their volume. However, with black holes, all the information seemed to be encoded on their surface.

This realization paved the way for Gerard ‘t Hooft and later Leonard Susskind to propose the holographic principle in the 1990s. They posited that the principle applied to black holes and the universe at large. Essentially, the idea suggests that all the information required to describe a volume of space—be it a room or an entire universe—could be encoded on the boundary of that volume.

This is analogous to a hologram. A hologram is a two-dimensional surface that, when illuminated correctly, projects a three-dimensional image. Similarly, the holographic principle proposes that our perceived 3D universe is a projection of information stored on a distant 2D boundary.

Holographic projection: 4biddenknowledge Inc

A pivotal development that bolstered the credibility of the holographic principle came from string theory, particularly with the discovery of the AdS/CFT correspondence by Juan Maldacena. This discovery demonstrated a concrete example wherein a gravitational theory in a certain space (AdS or Anti-de Sitter space) is equivalent to a quantum field theory without gravity on the boundary of that space. This correspondence provided a robust framework within which the holographic principle could be tested and explored.

So, what does this mean for our understanding of reality? If the holographic principle holds, our intuitive understanding of dimensional space is fundamentally altered. The 3D world we interact with is an emergent phenomenon from a deeper, underlying 2D reality. It's as if our entire universe is a grand projection emanating from a distant cosmic screen.

Beyond its philosophical implications, the holographic principle offers practical utility in resolving some of the biggest conundrums in theoretical physics. For instance, the incompatibility between general relativity (which describes gravity) and quantum mechanics (which describes the realm of the very small) has eluded physicists for decades. The holographic principle connects string theory and offers a possible path to a unified theory of quantum gravity.

In addition, the principle may hold answers to understanding the nature of entropy and information in the universe. By proposing that the universe's entropy is maximized when spread out on a bounding surface, it provides insights into the behavior of information in quantum systems. It has potential ramifications in areas like quantum computing.

It's worth noting, however, that while the holographic principle is a fascinating and potentially revolutionary concept, it remains on the frontier of theoretical physics. Empirical evidence directly supporting the principle is yet to be found. However, this hasn't deterred physicists from exploring its implications and pushing the boundaries of our understanding.

In summary, the holographic principle is one of modern physics's most captivating and enigmatic ideas. Proposing that our universe is but a holographic projection from a deeper 2D reality, it challenges our perceptions of space, reality, and information. As with many revolutionary ideas, it is enveloped in a shroud of mystery, awaiting the combined efforts of curious minds to unveil its truths. Whether it will stand the test of time and empirical scrutiny remains to be seen. Yet, its mere proposition, bridging the vast and the minute, has already expanded the horizons of theoretical physics.

The Basic Mathematics of Fractals Unveiling the Complex Patterns of Simplicity

Fractals are intricate mathematical structures that exhibit self-similarity across different scales. They are not just abstract mathematical concepts but also mirror many patterns found in nature, such as the branching of trees, the structure of snowflakes, and the ruggedness of coastlines. Understanding the basic mathematics of fractals is essential to appreciating their beauty and profound implications in various fields, from geometry and physics to computer science and art.

Understanding Fractals: Definition and Characteristics

A fractal is defined by three primary characteristics: self-similarity, infinite complexity, and fractional dimension. These properties make fractals a fascinating subject within mathematics and beyond.

1. **Self-Similarity:** Fractals exhibit self-similarity, meaning that their structure is made up of smaller copies of the whole. No matter how much you zoom into a fractal, you will see the same shape or pattern. This property can be exact or approximate but is a defining feature of fractals.

2. **Infinite Complexity:** Fractals are structures that exhibit complexity at every scale. Unlike traditional geometric shapes

that become simpler as you zoom in, fractals maintain or even increase their complexity the closer you look. This infinite complexity means that no matter how much you zoom into a fractal, you will always find new details.

3. **Fractional Dimension:** Perhaps the most counterintuitive aspect of fractals is their dimension. While we are familiar with the dimensions of lines (1D), squares (2D), and cubes (3D), fractals often have dimensions that are not whole numbers but fractions. This fractional dimension, or Hausdorff dimension, measures the complexity of the fractal—how much space it fills as you zoom in on it.

Key Mathematical Concepts in Fractals

When delving into the mathematics of fractals, several key concepts and tools are used to describe and generate these structures.

- **Iterative Processes:** Many fractals are generated through iterative processes, where a simple rule is applied repeatedly. The Mandelbrot set, one of the most famous fractals, is created by iterating the function shown below, where z and c are complex numbers. The behavior of this iteration, whether it converges or diverges, determines the points that belong to the Mandelbrot set.

$$z \to z^2 + c$$

- **Recursion:** Fractals often use recursive methods, where a procedure calls itself. This is seen in the construction of the

Sierpiński triangle, where a triangle is divided into smaller triangles, and the process is repeated for each smaller triangle. This recursive approach is fundamental to understanding how simple rules can lead to complex structures.

- **Scaling and Self-Affinity:** Fractals exhibit scaling properties, meaning that small portions of the fractal are related to the whole by a consistent ratio. The scaling factor quantifies this property and is crucial in determining the fractal dimension. Self-affinity is a related concept where the scaling may differ in different directions, as seen in fractals like the coastline, which is more irregular horizontally than vertically.

Calculating Fractal Dimension

The fractal dimension is a key quantitative measure of a fractal's complexity. It indicates how a fractal's detail changes with scale. There are several methods to calculate this, but the box-counting method is one of the most accessible.

- **Box-Counting Method:** This involves covering the fractal with boxes of a certain size and counting how many boxes contain a part of the fractal. As the size of the boxes decreases, the number of boxes needed to cover the fractal increases in a specific way.

The fractal dimension D is then calculated using the following formula where ϵ is the size of the boxes and $N(\epsilon)$ is the number of boxes containing part of the fractal.

$$D = \lim_{\epsilon \to 0} \frac{\log N(\epsilon)}{\log(1/\epsilon)}$$

- **Hausdorff Dimension:** For more mathematically rigorous work, the Hausdorff dimension is used. It involves a more complex measure that counts and weighs the boxes based on their size, providing a more accurate dimension, especially for irregular fractals.

Applications of Fractal Mathematics

The mathematics of fractals extends beyond theoretical work into numerous practical applications.

- **Image Compression:** Fractals can compress images efficiently because natural images often exhibit fractal-like patterns. The fractal compression technique identifies self-similar regions within an image and uses these to reconstruct the image from a smaller data set.

- **Modeling Natural Phenomena:** Many natural structures, from the branching of rivers to the shape of clouds, are fractal. Using fractal mathematics, scientists can model these structures more accurately than with traditional geometric forms, improving our understanding of natural processes and enhancing simulations in meteorology, geology, and ecology.

Fractals in nature: 4biddenknowledge Inc

- **Antenna Design:** Fractal shapes are used in the design of compact antennas. The fractal design allows for a broader frequency range and better signal reception within a smaller surface area, improving performance in various communication devices.

- **Art and Aesthetics:** Fractals have inspired artists and musicians, leading to creations that explore the beauty of mathematical patterns. The intricate designs of fractals have been used in everything from painting and sculpture to digital art and music composition.

Conclusion

The basic mathematics of fractals reveals a world where complexity arises from simplicity and infinite detail can be encoded in finite forms. This fascinating branch of mathematics deepens our understanding of the natural and digital worlds and challenges our perceptions of space, time, and beauty. Through fractals, we glimpse the underlying patterns that connect the universe, offering profound insights into the nature of reality and our place within it.

Linking Fractals and the Holographic Universe

The fields of theoretical physics and mathematics are filled with abstract ideas that constantly redefine our understanding of reality. Two such intriguing concepts are fractals and the holographic principle. On the surface, they might seem disparate, addressing different aspects of the universe. Yet, when examined closely, they intertwine in profound ways, suggesting that the universe's structure and behavior might be far more intricate and interconnected than previously believed.

Fractals: The Infinite Nested Reality

Fractals, as previously discussed, represent self-similar patterns across different scales. They reveal an endless web of complexity originating from simple rules. Nature, from the branching of trees to the formations of galaxies, offers a plethora of examples where fractal patterns are evident.

The essence of fractals suggests that the universe may be built upon iterative processes that manifest complexity from simplicity. The odds that we are living in base reality are very high. It is more likely that we are living inside of an ancestor universe.

The possibility that humanity and our universe might be the result of an ancestor-created universe suggests that what we perceive as reality could be a sophisticated simulation designed by an advanced civilization. This idea intersects with concepts from philosophy, physics, and computer science, challenging our understanding of existence and reality. Let's explore this hypothesis in depth.

The Simulation Hypothesis

The simulation hypothesis suggests that our reality might be an artificial construct created by a more advanced civilization. This idea has gained traction due to advancements in technology and theoretical physics. The hypothesis suggests that future civilizations might have the computational power to simulate entire universes, including the consciousness of beings within them.

Key Concepts Supporting the Hypothesis

- **Computational Advancements:** As our technology advances, we've seen significant improvements in computational power, virtual reality, and artificial intelligence. If this trend continues, it's conceivable that future civilizations could create highly sophisticated simulations indistinguishable from what we perceive as reality.

- **Moore's Law:** Moore's Law, which states that the number of transistors on a microchip doubles approximately every two years, implies exponential growth in computational capacity. While this trend may not continue indefinitely, it suggests that advanced civilizations could achieve computational feats beyond our current imagination.

- **Quantum Mechanics:** Quantum mechanics, with its probabilistic nature and observer effect, hints that reality may not be as concrete as it seems. Some interpretations of quantum mechanics suggest that reality is information-based and that our universe might function like a giant quantum computer.

Philosophical Considerations

- **Descartes' Evil Demon**: René Descartes hypothesized that an evil demon could deceive him about the nature of reality. Similarly, the simulation hypothesis suggests that an advanced civilization could control or create our perceptions of reality.

- **Brain in a Vat:** This thought experiment posits that a brain kept alive in a vat and fed false sensory information could experience a completely artificial reality. The simulation hypothesis extends this idea to our entire universe.

Scientific and Technological Implications

- **Nested Simulations:** If our universe is a simulation, the creators of our simulation may be themselves simulated beings created by an even more advanced civilization. This leads to the concept of nested simulations or a hierarchy of simulated realities.

- **Detection of Simulations:** Some scientists and researchers propose ways to test if we are living in a simulation. This might involve detecting anomalies or "glitches" in the fabric of reality, similar to bugs in a computer program.

Ethical and Existential Implications

- **Purpose and Meaning:** If we are in a simulation, it raises questions about the purpose of our existence. Are we part of an experiment, a form of entertainment, or a historical recreation by our ancestors? This can influence our understanding of meaning and purpose in life.

- **Free Will and Determinism:** The simulation hypothesis might imply that the simulation's programming predetermines our actions, challenging traditional notions of free will. Alternatively, the creators could have designed the simulation to allow for free will within certain parameters.

- **Moral Responsibility:** If we create our own simulations with conscious beings, we must consider our ethical responsibilities toward them. Similarly, if we are in a simulation, we must ponder the moral obligations of our creators toward us.

Potential Evidence and Arguments

- **Mathematical Nature of Reality:** Some scientists, like physicist Max Tegmark, argue that the mathematical nature of the universe suggests it could be a simulation. The precise mathematical constants and the fine-tuning of physical laws could indicate an underlying computational framework.

- **Complexity and Simplicity:** The fractal holographic model suggests that complexity arises from simplicity through recursive patterns. This mirrors how simulations work, with simple rules generating complex behaviors, reinforcing the idea that our universe might be a computational construct.

Counterarguments and Challenges

- **Technological Limitations:** While future civilizations might achieve incredible computational power, there are still unknown physical and computational limits. The energy and resources required to simulate entire universes might be prohibitive.

- **Philosophical Rejections:** Some philosophers argue that the simulation hypothesis is unfalsifiable and not scientifically useful. It may also lead to solipsism, the idea that only one's mind is sure to exist, which is generally rejected in philosophical discourse.

- **Complexity of Consciousness:** Simulating consciousness and subjective experiences is an enormous challenge. The nature of consciousness remains one of the biggest mysteries in science, and creating a realistic simulation might be far more difficult than simulating physical phenomena.

The hypothesis that humanity and our universe might be the result of an ancestor-created universe presents a fascinating and provocative view of reality. It integrates advanced computational theories, quantum mechanics, and philosophical inquiries, challenging our understanding of existence and consciousness. While compelling arguments and potential evidence support this idea, it also faces significant scientific, philosophical, and ethical challenges. Whether or not we live in a base reality, exploring this hypothesis encourages us to reflect on the nature of reality, our place in the cosmos, and the potential future of human civilization.

The Holographic Principle: A Boundary of Information

On the other hand, the holographic principle posits that the information contained within a particular volume of space can be entirely represented on the boundary of that space. This challenges our traditional perception of dimensionality, hinting at a universe where our three-dimensional reality is a projection from a two-dimensional boundary.

Bridging the Concepts

Now, how do these two concepts intersect?

- **Nature of Information and Structure:** Both fractals and the holographic principle fundamentally deal with information and structure. Fractals show that infinite complexity can arise from simple iterative processes. Similarly, the holographic principle suggests that vast amounts of information in a three-dimensional space are encoded in a two-dimensional boundary. Both challenge the conventional understanding of space and information density.

- **Scale and the Universe's Fabric**: Fractals operate on the principle of self-similarity across scales. While primarily concerned with encoding information on a boundary, the holographic principle also touches upon the idea of scale, especially when considering how a lower-dimensional boundary can represent a higher-dimensional space. There's a shared theme of depth beneath the surface and an intrinsic connection between the micro and macro scales of the universe.

As Above So Below: 4biddenknowledge Inc

- **Quantum Gravity and the Universe's Geometry:** One of the primary motivations behind exploring the holographic principle is to understand quantum gravity, a theory that reconciles quantum mechanics and general relativity. String theory, which delves deeply into the holographic principle, also touches upon the fractal nature of space-time at the tiniest scales, suggesting a possible fractal holographic relationship in the very fabric of the universe.

- **Cosmic Fractals and Holograms:** If we take a step back and look at the universe's large-scale structure—galaxies grouped into clusters, which in turn form superclusters, creating a vast cosmic web—we can observe a fractal pattern. Now, if this vast cosmic structure were to be a holographic projection from a cosmic horizon (as the holographic principle might suggest), the universe would become a grand fractal hologram.

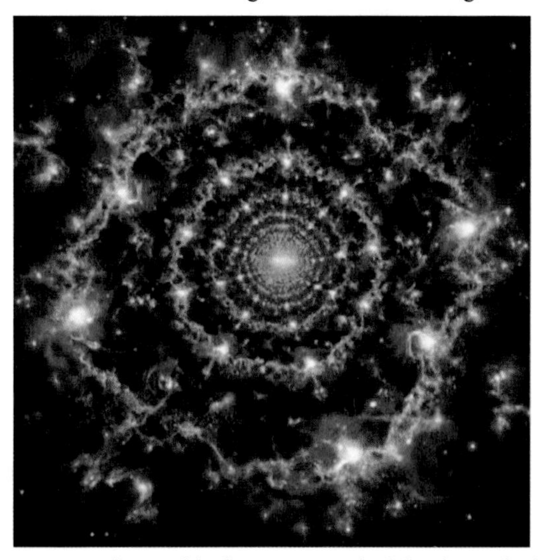

Universe as a fractal hologram – 4biddenknowledge Inc

- **Consciousness and Reality Perception:** On a more speculative note, if our consciousness is also a result of the universe's fractal and holographic nature, our perception of reality becomes an emergent phenomenon from the underlying patterns and encodings. Both theories could work in tandem to shape how reality unfolds and is perceived.

Implications and the Grand Symphony

When we imagine a universe governed both by fractal geometry and holographic projections, we envision a cosmos of unimaginable beauty and complexity. Such a universe would be endlessly intricate, where every part reflects the whole in a dance of patterns and projections.

From a more practical perspective, intertwining fractals and the holographic principle could advance our understanding of fundamental physics. Theories of quantum gravity, the nature of black holes, the Big Bang, and even the eventual fate of our universe might be better understood at the intersection of these ideas.

Furthermore, every bit of the universe, from the tiniest quantum fluctuation to the vast expanse of cosmic structures, is interconnected in this proposed paradigm. This brings forth a holistic view of the cosmos, where understanding one aspect, be it through fractals or holography, provides insights into the other.

Conclusion

While we are still at the frontier of understanding the true nature of our universe, and many questions remain unanswered, the melding of ideas like fractals and the holographic principle presents a tantalizing avenue of exploration. Such interdisciplinary intersections might hold the key to unlocking the universe's most profound mysteries, revealing a cosmos that's both a fractal and a hologram—a grand symphony of patterns and projections.

Chapter 2

Origins of Fractal Thinking

Fractal thinking, as understood in the modern context, emerges from the study of self-similar structures across different scales. But the seeds of this mode of thought can be traced back to ancient civilizations, philosophical discourses, and a long-standing human fascination with patterns in nature. While the term "fractal" and its formal mathematical understanding is relatively recent, the intuition behind it spans millennia.

Ancient Civilizations and Patterns

Humans have always been observers of nature, and the patterns found in the natural world have never ceased to captivate the human imagination. The branching of trees, the veins in leaves, river deltas, and erosion patterns are just a few examples of naturally occurring structures that carry a fractal essence. Ancient civilizations, from the Egyptians to the Mayans, incorporated repetitive, pattern-based designs in their architecture, art, and textiles. While they might not have consciously recognized these as "fractals," the essence of fractal thinking was embedded in their reverence for natural patterns.

Ancient artisans used fractal patterns – 4biddenknowledge Inc

Philosophical Underpinnings

Philosophers have long contemplated the nature of infinity, divisibility, and the microcosm-macrocosm relationship. Ancient Greek thinkers pondered the infinite divisibility of matter. Zeno of Elea, in his paradoxes, touched upon ideas that resonate with fractal concepts. His paradoxes deal with infinite division, suggesting that motion is an illusion because a moving object never reaches its destination as it's always halving the distance. This infinite iterative process is reminiscent of fractal structures.

Similarly, the concept of "as above, so below," rooted in Hermeticism, reflects that the patterns and structures observed at one level of the cosmos can be replicated at other levels. This is a cornerstone of fractal thinking, emphasizing self-similarity across scales.

As Above So Below fractal patterns – 4biddenknowledge Inc

Cultural Reflections

Various cultures around the world have also echoed fractal thinking in their traditional designs and artworks. African architecture, for instance, exhibits fractal patterns, from the layout of villages to the design of individual buildings. Similarly, in Japanese culture, self-similar patterns can be found in certain artworks, textiles, and even in the concept of the infinite reflection between two mirrors.

Scientific Precursors

Before the formal introduction of fractal geometry, several scientists and mathematicians had touched upon ideas that align with fractal thinking. The coastline paradox, introduced by Lewis Fry Richardson, is a classic example. It highlights that the length of a coastline seems to increase indefinitely as the measurement scale becomes smaller. This observation is now recognized as a classic fractal phenomenon.

Furthermore, mathematicians like Karl Weierstrass and Georg Cantor, in the late nineteenth and early twentieth centuries, developed functions and sets that had properties resonating with fractal structures, even though they didn't use the term "fractal" or see them in that specific light.

Benoît B. Mandelbrot and the Dawn of Fractal Geometry

It was only in the twentieth century, with the advent of computers and the work of Benoît B. Mandelbrot, that the term "fractal" was coined, and the concept was rigorously explored. By leveraging modern computers' computational power, Mandelbrot delved into patterns that seemed irregular and chaotic. He introduced the term "fractal" to describe shapes that look similar (or equally "rough") no matter how much you zoom in. His work on the Mandelbrot set, a set defined by a simple iterative formula but producing infinite complexity, became the iconic image of fractal geometry.

Mandelbrot's work linked diverse fields and phenomena, from the fluctuations of financial markets to the structure of galaxies, under the umbrella of fractal geometry. His book, The Fractal Geometry of Nature, published in 1982, cemented the idea that the seemingly irregular shapes and patterns in nature could be described using fractal geometry.

Wikimedia Commons - Benoît B. Mandelbrot

Conclusion

Fractal thinking is a testament to humanity's long-standing quest to understand the patterns that underpin our universe. From ancient art and architecture to modern computational mathematics, the journey of fractal thinking spans cultures, disciplines, and epochs. It embodies the human spirit's perennial quest to discern order in chaos, to find unity in diversity, and to unravel the intricate dance of simplicity and complexity in the fabric of existence.

Today, as we stand at the intersection of science, art, and philosophy, fractal thinking offers a lens to view the world with a renewed sense of wonder, tracing patterns that bridge the infinitesimally small to the astronomically large. The origins of fractal thinking, rooted in ancient observations and refined through modern mathematics, remind us that the universe's elegance and intricacy have always been within our grasp, waiting to be discovered and understood.

Historical Perspective of Fractals: From Ancient Mosaics to Mandelbrot

Though a concept solidified in the latter half of the twentieth century, Fractals have roots stretching deep into humanity's artistic and intellectual history. The evolution of fractal thinking, from the intricate designs of ancient mosaics to the computational explorations of Benoît Mandelbrot, underscores the innate human attraction to patterns and self-similarity.

Ancient Recognition

Long before the term "fractal" was coined, ancient civilizations created designs and structures that embodied the principle of repeating patterns on different scales. Consider the ancient Roman and Byzantine mosaics, where tessellated patterns exhibit what we recognize today as fractal-like properties. These decorative arts, consisting of small, geometrically arranged pieces of stone or glass, often displayed patterns that had a degree of self-similarity, especially in religious and culturally significant designs.

Similarly, in India, the recursive patterns in temple architecture and traditional art forms like rangoli (artistic designs made on floors using colored rice, sand, or flower petals) reflected an intuitive grasp of fractal structures. Such patterns, revered for their beauty and symmetry, formed a bridge between the spiritual and the artistic.

Medieval Iterations

The medieval era, particularly in Islamic and Gothic architecture, saw a flourishing of design principles that are distinctly fractal in nature. Islamic art, for instance, given its prohibition on depicting living figures, leaned heavily on geometric patterns. The intricate designs in mosques, often consisting of tiles placed in repetitive and nested patterns, demonstrate a clear fractal essence.

Gothic cathedrals of Europe, with their repeating arches, spires, and ornate windows, also reflect a preoccupation with self-similar design, albeit achieved more through architectural forms than two-dimensional patterns.

Renaissance Reflections

The Renaissance period, known for its profound appreciation of art and science, subtly integrated fractal-like structures in various works. The branching patterns in Leonardo da Vinci's sketches, the recursive nature of motifs in tapestries, and the layered perspectives in paintings all hint at an implicit understanding of fractal geometry.

Philosophical Foundations

In parallel to artistic manifestations, philosophical inquiries during the Enlightenment period began to touch upon concepts that resonate with fractal thinking. The notion of the infinitely divisible, pondered upon by thinkers like Immanuel Kant and Bernard Bolzano, touched upon the challenge of reconciling the finite and the infinite—a central theme in fractal structures.

Precursors in Modern Mathematics

Before Mandelbrot's crystallization of fractal geometry, the mathematical world had already seen glimpses of fractal-like structures. During the nineteenth and early twentieth centuries, mathematicians such as Karl Weierstrass and Georg Cantor developed functions and sets that displayed properties akin to fractals. Cantor's set, for instance, created by repeatedly removing middle thirds from a line segment, presented a structure with an infinitely intricate boundary—very much in line with fractal characteristics.

Helge von Koch's snowflake curve, described in 1904, provided another early example. Starting with an equilateral triangle and successively adding smaller triangles at each stage, the curve exhibits clear self-similarity yet possesses a finite area with an infinite perimeter.

Mandelbrot's Revolution

The term "fractal" and a comprehensive exploration of its properties entered the scientific lexicon mainly due to the efforts of Benoît B. Mandelbrot. With the advantage of 20th-century computational power, Mandelbrot began to delve into irregular and chaotic patterns, searching for an underlying order.

His discovery of the Mandelbrot set, a set of points in the complex plane with strikingly beautiful and endlessly intricate patterns emerging from simple iterative processes, brought the concept of fractals to the forefront of mathematical research. His seminal book, The Fractal Geometry of Nature, published in 1982, championed the idea that many natural phenomena, previously considered "messy" or "irregular," could be effectively described using fractal geometry.

Conclusion

From the designs etched in ancient stones to the computational patterns generated in modern computers, the journey of fractals provides a compelling narrative of humanity's enduring engagement with patterns, complexity, and beauty. It's a testament to the fact that concepts, even if not named or formally recognized, can thread through centuries, waiting for the right moment of crystallization.

The history of fractals, bridging art, philosophy, and science, reminds us of the interconnected tapestry of human knowledge. It underscores that even in the face of the infinite and the intricate, humanity has always sought and often found underlying patterns and symmetries. The story of fractals, from ancient mosaics to Mandelbrot, is a journey from seeing patterns in the world to understanding the very fabric of the universe.

Fractal Geometry in Nature: Ferns, Coastlines, and Galaxies

In its myriad forms and scales, nature displays a level of complexity that has long intrigued scientists, mathematicians, and artists alike. One of the most fascinating realizations of the last century has been recognizing that many of these complexities can be understood and described using fractal geometry. This geometry, different from classical Euclidean geometry, has shown that seemingly irregular and chaotic forms, from the microscopic to the cosmic, often follow a pattern of self-similarity across scales. Ferns, coastlines, and galaxies provide compelling examples of fractal geometry at work among the vast array of natural phenomena.

Fractal Geometry in Nature: Ferns, Coastlines, and Galaxies – 4biddenknowledge Inc

Ferns: The Delicate Symmetry

If one has ever taken a stroll through a dense forest or a botanical garden, the delicate and intricate structure of ferns is hard to miss. At a casual glance, the fern's fronds reveal a branching pattern. On closer inspection, each leaflet (or pinna) of a frond seems to resemble a smaller version of the entire frond itself. Dive deeper; even smaller leaflets (pinnules) further reflect this recursive branching.

This self-similarity, where parts of the fern echo the structure of the whole, is a classic representation of fractal geometry in the botanical world. The growth pattern of ferns can be modeled using iterative processes and simple mathematical rules, resulting in surprisingly accurate representations of real-world ferns. These models capture the fern's aesthetics and provide insights into the evolutionary advantages such a growth pattern might confer, like maximizing sunlight exposure.

Coastlines: The Paradox of Length

One of the earliest and most famous real-world applications of fractal thinking came with the coastline paradox. It's a simple yet profound observation: the measured length of a coastline depends on the size of the measuring stick. As one uses a smaller and smaller ruler to measure, the length seems to increase, accounting for all the nooks and crannies. In the limit, as the ruler's size approaches zero, the coastline appears infinitely long!

This counterintuitive result can be understood through fractal geometry. Coastlines, like many natural boundaries, display self-similar patterns. The irregularities and patterns observed at hundreds of kilometers persist even at scales of meters or centimeters. Thus, the "roughness" or "complexity" of the coastline remains consistent across scales, making it a fractal entity. While accepting an infinite length for real-world coastlines is conceptually challenging, the fractal description provides tools to characterize their complexity. It offers insights into erosional patterns and geological processes.

Galaxies: Cosmic Webs

Moving from terrestrial scales to the vast cosmic expanses, fractal patterns emerge yet again in the distribution of galaxies. For a long time, the universe, on large scales, was thought to be homogeneous. However, as observational technology advanced, a more intricate picture surfaced. Galaxies weren't just randomly sprinkled throughout space; they formed clusters, superclusters, and intricate networks known as cosmic webs.

These cosmic structures, when analyzed, showcased fractal characteristics. The distribution of galaxies in certain regions of the universe displayed self-similarity over a range of scales. While this doesn't imply the entire universe is a fractal, it does suggest that parts of it can be effectively described using fractal geometry.

Understanding the universe through the lens of fractals also ties into theories about its formation. With minute fluctuations in density, the early universe went through gravitational coalescence, leading to the filamentary structures of galaxies and voids we observe today. Fractal geometry offers tools to model, describe, and understand these processes and patterns.

Fractal geometry, in capturing the essence of self-similarity and recursion, provides a powerful paradigm for understanding the natural world. From the leaves rustling underfoot to the distant twinkling galaxies, the echo of fractals is pervasive. Recognizing these patterns is more than a mere mathematical exercise; it's a testament to our universe's interconnectedness and inherent patterns.

The study of fractals in nature—whether in ferns, coastlines, or galaxies—reiterates a profound realization: that the laws and patterns governing the microscopic and the cosmic are interwoven, bridging the immense chasm of scales with elegance and simplicity. In celebrating fractals, we celebrate nature's propensity for complexity and order, a dance of chaos and pattern that paints the tapestry of existence.

Human Beings and Programming Codes: A Detailed Exploration

Social and Cultural Conditioning

Humans are heavily influenced by the environment and society they grow up in. This influence can be seen as a form of programming where beliefs, behaviors, and norms are instilled in individuals from a young age.

Early Childhood Programming

- **Family and Caregivers:** The initial "programming" starts with parents and caregivers, who teach children fundamental behaviors and beliefs through direct instruction and modeled behavior.

- **Language and Communication:** The language we learn and the way we communicate are forms of cultural coding. Our words and phrases shape our perception of reality and influence our thoughts.

Education System

- **Curriculum and Pedagogy:** The education system further programs individuals by providing structured knowledge and societal norms. It often emphasizes certain viewpoints while excluding others, shaping how individuals think and understand the world.

- **Social Interactions:** School environments also program social behaviors, teaching norms, and values through peer interactions and institutional rules.

Media and Technology

- **Mass Media:** Television, movies, news, and advertising continuously program beliefs and desires by portraying certain lifestyles, values, and ideologies as desirable or normal.

- **Social Media:** Algorithms on social media platforms reinforce existing beliefs and behaviors by showing content that aligns with users' preferences, creating echo chambers that amplify certain viewpoints.

Biological and Biochemical Programming

The human body operates on a set of biological and biochemical codes, often referred to as genetic and hormonal programming.

Genetic Programming

- **DNA and Genes:** Our genetic code, made up of DNA, dictates our physical characteristics and predispositions to certain behaviors and health conditions. Genes are turned on and off through epigenetic mechanisms, which can be influenced by environmental factors.

Hormonal Influence

- **Endocrine System:** Hormones act as chemical messengers, influencing a wide range of bodily functions and behaviors.

- **Cortisol:** Often called the stress hormone, influences how we respond to stress and can affect mood, energy levels, and immune function.

- **Oxytocin:** Known as the bonding hormone, plays a role in social bonding, sexual reproduction, and during and after childbirth.

- **Testosterone and Estrogen:** These sex hormones influence a variety of behaviors, including aggression, sexual desire, and social interaction patterns.

Neurotransmitters

- **Dopamine:** Often referred to as the pleasure chemical, it plays a critical role in reward-motivated behavior. It reinforces behaviors that provide pleasure and satisfaction.

- **Serotonin:** This neurotransmitter is crucial to mood regulation. Imbalances can lead to conditions like depression and anxiety.

- **Endorphins:** These are the body's natural painkillers and mood elevators, often released during exercise, excitement, pain, and consumption of spicy food.

Psychological Programming

Psychological factors and cognitive processes also shape human thoughts and behaviors.

Cognitive Biases

- **Confirmation Bias:** The tendency to search for, interpret, and remember information that confirms pre-existing beliefs.

- **Anchoring Bias:** Relying too heavily on the first piece of information encountered (the "anchor") when making decisions.

Behavioral Conditioning

- **Classical Conditioning:** Learning through association, where a neutral stimulus becomes associated with a meaningful stimulus, eliciting a conditioned response.

- **Operant Conditioning:** Learning through rewards and punishments, where behaviors are reinforced or discouraged based on the consequences.

Social Learning

- **Modeling and Imitation:** Individuals learn behaviors by observing and imitating others, especially role models and authority figures.

The Integration of Codes

Human beings operate through a complex integration of social, biological, biochemical, and psychological programming codes.

Interaction of Influences

- **Biopsychosocial Model:** This model suggests that biological, psychological, and social factors play a significant role in human functioning.

- **Feedback Loops:** There are continuous feedback loops between these factors. For instance, stress (psychological) can influence cortisol levels (biological), which in turn can affect behavior (social).

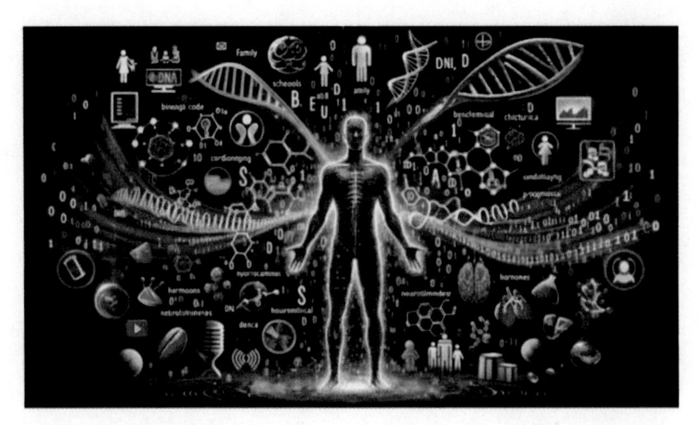

Human being receiving programming codes – 4biddenknowledge Inc

Adaptation and Flexibility

- **Neuroplasticity:** The brain's ability to reorganize itself by forming new neural connections throughout life. This flexibility means that while various codes program humans, they also have the capacity to change and adapt.

- **Learning and Experience:** Lifelong learning and new experiences can alter existing programming, leading to personal growth and change.

Conclusion

Human beings operate off a multitude of programming codes that encompass social, cultural, biological, biochemical, and psychological influences. These codes shape our beliefs, behaviors, and perceptions of reality. Understanding these programming codes can provide insights into human behavior and the potential for personal and societal change.

Mathematical Wonders of Fractals: The Birth of Fractal Geometry

In the realm of mathematics, few ideas are as captivating and aesthetically awe-inspiring as fractals. These structures, defined by self-similarity and intricate detail at every scale, have revolutionized how we view geometry and bridge the gap between mathematics, nature, and art. The birth of fractal geometry marked a transformative period in the history of mathematics, presenting tools to delve into the complexity of seemingly chaotic systems.

Fractals: Beyond Traditional Geometry

Traditional geometry, the one many encounter in early school years, is dominated by familiar shapes: lines, circles, triangles, and the like. These shapes, derived from Euclidean principles, work remarkably well for many applications. However, they fall short when describing the raggedness of a coastline, the irregular branching of trees, or the spirals of galaxies.

Enter fractal geometry, which challenges these classical views. Unlike Euclidean geometry's smooth surfaces and curves, fractals are defined by rough, self-replicating patterns. The name "fractal" itself, derived from the Latin word "fractus," meaning "broken," hints at this inherent irregularity.

Hallmarks of Fractal Geometry

At the heart of fractal geometry lies the concept of self-similarity. This means that a small portion of the fractal, when magnified, mirrors the structure of the larger whole. This property persists across scales, whether you zoom in or zoom out.

Another defining characteristic of fractals is their non-integer or fractional dimension. Unlike a line, which is one-dimensional, or a square, which is two-dimensional, fractals exist in between these clear-cut dimensions. This can be counterintuitive, but it provides a way to measure the "complexity" or "roughness" of fractals.

The Pioneers and Their Constructs

Long before the term "fractal" was introduced, mathematicians unwittingly explored fractal-like structures. One of the early instances is the Cantor set, proposed by Georg Cantor in the late nineteenth century. By repeatedly removing middle thirds from a line segment, Cantor created a set with a counterintuitive property: it had zero length but was uncountably infinite in points!

Wikimedia Commons - Georg Cantor

In the early 20th century, Helge von Koch presented the Koch curve (or Koch snowflake), a simple yet infinite curve constructed by adding triangles at each iteration. Similarly, the Sierpiński triangle, a structure formed by repeatedly removing equilateral triangles from a larger one, showcased self-similarity and an intricate pattern arising from simple rules.

Mandelbrot: The Prophet of Fractal Geometry

While these early constructs hinted at the beauty and depth of fractals, it was Benoît B. Mandelbrot who truly paved the way for the formalization of fractal geometry. In the 1970s, armed with computational tools and an innate curiosity to explore the irregularities of nature, Mandelbrot began his pioneering work.

His most iconic discovery is the Mandelbrot set, a set of complex numbers that, when visualized, creates mesmerizing patterns filled with whirlpools, spirals, and miniature copies of the whole set. The simple iterative formula at its core, contrasted with its infinitely intricate output, embodies the magic of fractal geometry.

Mandelbrot's seminal work, The Fractal Geometry of Nature, published in 1982, was a clarion call for the appreciation and application of fractals. He posited that the irregularities in nature often dismissed as anomalies or noise, were genuinely part of the grander pattern. They weren't just to be acknowledged; they were to be celebrated.

Applications and Beyond

The birth of fractal geometry had profound implications beyond pure mathematics. Today, fractals find applications in diverse fields like computer graphics, where they generate realistic landscapes and terrains; in biology, for modeling patterns like blood vessels and plants; in finance, for capturing the unpredictable movements of stock markets; and even in telecommunications, where they aid in optimizing antenna designs.

Beyond their utility, fractals serve as a humbling reminder of the wonders of mathematics. They challenge our intuitive notions of dimension, highlight the intricate patterns underlying chaotic systems, and showcase the endless beauty that can arise from simple rules.

Conclusion

The birth of fractal geometry, in redefining the boundaries of mathematical exploration, has unveiled a universe of patterns, chaos, and wonder. It's a testament to the power of mathematical abstraction to illuminate the hidden patterns of our world.

We see the dance of numbers, formulas, and nature in fractals. They remind us that mathematics, often viewed as a cold and logical discipline, can be a wellspring of beauty, mirroring the intricacies and wonders of the universe. From the branching of trees to the swirls of galaxies, the mathematical wonders of fractals echo the timeless words of Galileo: "The universe cannot be read until we have learned the language... It is written in mathematical language."

Chapter 3

The Holographic Principle: A Primer

Basic Understanding: Every Part Contains the Whole Within Fractals

AT ITS CORE, THE concept of fractals challenges our traditional notions of scale and self-similarity. In a world where we often understand things in terms of their constituent parts, fractals present an intriguing reversal: every part contains an echo or a representation of the whole. This might sound paradoxical at first, but delving into the world of fractals can reveal deep insights into the nature of these mathematical structures, patterns, and systems in the world around us.

Defining Fractals

Fractals are complex structures built from simple repetitions. They're mathematical sets that exhibit self-similarity, meaning they appear similar at any level of magnification. This property results in intriguing patterns where a small section of the fractal can be expanded, revealing a pattern nearly identical to the larger picture.

Every Part Reflects the Whole

Imagine a tree. At first glance, it might seem like a straightforward structure with a trunk that branches out. However, on closer examination, one can observe that each branch is like a smaller tree with sub-branches. These sub-branches further split, resembling the branch from which they sprouted, and so on. In this way, a single branch or even a sub-branch can provide insights into the structure and form of the entire tree.

Holographic fractals – 4biddenknowledge Inc

The Romanesco broccoli provides another vivid example from nature. This vegetable's fractal nature is apparent in its cone-shaped structures. Each cone comprises smaller cones that resemble the larger ones, and this pattern continues to an almost microscopic level. Such examples are not just coincidences of nature but reflections of a principle intrinsic to fractals: the whole can be understood by examining its parts, and vice versa.

Implications Beyond Geometry

This concept has philosophical and practical implications extending far beyond geometry and mathematics. In many world philosophies and spiritual traditions, the idea that every part of a system contains the essence of the whole is a recurring theme. This belief often underscores the interconnectedness of all things and the inherent patterns that govern systems, whether they're biological, they're or cosmic.

In practical terms, understanding this fractal property has led to breakthroughs in fields as varied as medicine, telecommunications, and computer graphics. For example, in medicine, understanding the fractal nature of blood vessels or bronchial trees in the lungs has aided in developing improved diagnostic tools and treatments. In telecommunications, fractal-shaped antennas, which capitalize on this self-similarity, can receive a broader range of frequencies.

Fractals and the Universe

One of the more profound contemplations regarding fractals revolves around the nature of our universe. Some scientists and philosophers propose that the universe might be fractal in nature. Galaxy clusters might arrange themselves in patterns that echo fractal mathematics. This notion suggests that we might gain insights into the universe's vast expanse by understanding smaller cosmic structures.

Moreover, the idea that every part contains the whole challenges our perception of individuality and separateness. If we apply this principle universally, it suggests that every individual entity, be it a person, planet, or star, contains the essence of the entire universe within it. It's a thought that resonates with many spiritual traditions, emphasizing unity and the interconnectedness of all beings.

Challenges in Understanding Fractals

While the principle of self-similarity in fractals might seem straightforward, truly grasping it can be counterintuitive. Euclidean geometry, with its lines, circles, and polygons, has dominated our understanding of space and form for centuries. Fractals challenge this linear perspective with their infinite complexity and scale variance. They don't fit neatly into one-dimensional, two-dimensional, or even three-dimensional categories. In fact, fractals often exhibit non-integer or fractional dimensions, which can be a challenging concept even for seasoned mathematicians.

Conclusion

The concept that every part contains the whole within fractals offers a fresh lens through which we can view the world. This perspective invites us to appreciate the intricate details, patterns, and structures that might otherwise be overlooked. It reminds us that the vastness of the universe and the intricate details of subatomic particles are part of a continuum bound by patterns and principles that echo across scales.

Fractals challenge us to see the beauty in complexity, appreciate the patterns that emerge from simplicity, and recognize that understanding a part can shed light on the vastness of the whole. Whether it's the branching of a tree, the spirals of a galaxy, or the patterns in a snowflake, the universe seems to whisper the secrets of its grand design in the language of fractals. And in this language, every part sings the song of the whole.

Black Holes and the Origins of the Holographic Universe Theory

The enigmatic nature of black holes has been a focal point of theoretical physics for decades. These astronomical entities, characterized by gravitational forces so intense that not even light can escape their grasp, have challenged our understanding of the cosmos and given rise to some groundbreaking theories, including the concept of a holographic universe.

Understanding Black Holes

Before diving into the holographic universe theory, one must first understand the basics of black holes. Born from the remnants of massive stars after they exhaust their nuclear fuel, black holes are regions of space-time where gravitational forces are overpowering. The "point of no return" around a black hole, where the gravitational pull becomes irreversibly strong, is known as the event horizon.

Black holes pose intriguing questions for physicists. At their center, according to classical theory, lies the singularity—a point where the gravitational force is infinite, and space-time curvature becomes extreme.

This idea challenges the very fabric of our understanding, prompting scientists to consider alternative and sometimes radical explanations.

The Information Paradox

Physicist Stephen Hawking posed a significant puzzle related to black holes in the 1970s. He proposed that black holes could emit radiation, now known as Hawking radiation. Over time, this radiation would cause the black hole to lose mass and eventually evaporate. The paradox arises when considering what happens to the information about particles that have fallen into the black hole once the black hole disappears. According to the principles of quantum mechanics, information cannot be destroyed. So, where does it go?

Hawking Radiation – 4biddenknowledge Inc

The information paradox has haunted physicists for years, prompting them to look for innovative solutions that might bridge the gap between general relativity (which describes the gravitational behavior of large objects) and quantum mechanics (which explains the dynamics of the smallest particles).

The Holographic Principle

It was within this context of understanding black holes and resolving the information paradox that the holographic principle was born. Proposed in the 1990s by physicist Gerard 't Hooft and later refined by Leonard Susskind, the holographic principle suggests that data on its boundary can describe the information contained within a volume of space. In other words, our three-dimensional reality might just be a projection of information stored on a distant, two-dimensional surface.

This idea can be visualized by thinking of a hologram on a credit card. The hologram is a two-dimensional surface, but it appears three-dimensional to our eyes. Similarly, the holographic principle postulates that the universe's entirety can be seen as a two-dimensional information structure "painted" on the cosmological horizon.

Black Holes as Holograms

To address the black hole information paradox, the holographic principle suggests that the information about particles falling into a black hole is not lost but rather stored on the event horizon's surface. Consequently, black holes can be envisioned as holograms, with their three-dimensional interior being a projection of information on their two-dimensional surface.

Recent studies using the framework of string theory have lent support to this idea. Researchers have found that the entropy, or the measure of disorder, of certain hypothetical black holes, matches up with the entropy of a lower-dimensional quantum system with no gravity. This alignment is seen as a significant step in reconciling general relativity with quantum mechanics.

Implications for the Grand Design of the Cosmos

If the holographic principle holds true for black holes, might it also apply to the universe? Some physicists believe so. They propose that the entire universe, with all its particles and fields, arises from information encoded on a boundary that exists in a higher-dimensional space.

This perspective offers a fresh approach to understanding the fabric of reality. Instead of seeing the universe as a three-dimensional space filled with stuff, we could see it as a hologram emerging from fundamental bits of information encoded on a distant two-dimensional surface.

Challenges and the Road Ahead

The holographic universe theory, while intriguing, remains highly speculative. More evidence and a complete theoretical framework are needed before this idea can be fully embraced. However, its potential implications are staggering, suggesting a universe vastly different from our perceptions.

Conclusion

Black holes, with their enigmatic nature, have pushed the boundaries of our understanding, leading to the proposal of groundbreaking theories like the holographic universe. These theories, rooted in the quest to solve the mysteries posed by black holes, might one day reshape our understanding of reality itself.

In the intertwining dance of light and gravity, the smallest particles and the vast cosmos, we find the eternal quest to understand the universe's true nature.

Fractals in the Cosmos: Implications for the Grand Design

Fractals, with their intriguing patterns of infinite self-similarity, have always captivated mathematicians and nature enthusiasts alike. Yet, their importance is not confined to their mesmerizing aesthetics. By virtue of their intrinsic properties, these mathematical structures may hold vital clues about the very fabric and grand design of the cosmos.

The Nature of Fractals

At their core, fractals are patterns that repeat at varying scales, both diminishing and expanding. This property of self-similarity means that a small part of a fractal when magnified, bears a striking resemblance to a larger section or even the whole. Fractals are not merely theoretical constructs; they manifest everywhere in nature, from the branching of trees and the spirals of galaxies to the intricate designs of snowflakes and the jaggedness of coastlines.

Fractals in Cosmic Structures

When we gaze at the vast expanse of the universe, it is astounding to recognize fractal patterns unfolding on an unimaginably grand scale. Galaxies, for instance, do not scatter randomly. They cluster in a manner that exhibits self-similarity across various scales. This means that a small cluster of galaxies can offer insights into the larger structure of the universe.

Star-forming regions within galaxies also follow fractal patterns. The distribution of matter, shaped by the interplay of gravity, radiation, and dark matter, frequently results in structures that resonate with fractal mathematics.

Implications for the Universe's Origin and Evolution

The presence of fractals across the cosmos prompts us to reconsider our understanding of the universe's origin and evolution. If the universe follows a fractal pattern, it suggests a level of order and predictability amid the vast chaos.

- **Origin:** The Big Bang theory posits a singular explosion from which the universe expanded. In a fractal universe, this explosion could be visualized not just as a chaotic scattering of matter but as an event that gave rise to patterns that would repeat and refine themselves over billions of years.

- **Evolution:** The cosmic web, with its vast voids and filamentous structures of galaxies, can be better understood through fractals. As the universe evolved, matter didn't just clump randomly. It followed patterns, much like a fractal evolving and replicating its designs across scales.

Dark Matter and Fractals

One of the most significant mysteries of modern cosmology is the enigmatic dark matter. This unseen and unidentified form of matter doesn't emit, absorb, or reflect light, but it exerts gravitational forces.

The filamentous structures that dark matter forms in the universe exhibit fractal patterns. Understanding these patterns might provide crucial insights into the nature and behavior of dark matter.

Fractals and Quantum Mechanics

On the opposite end of the scale, in the realm of the extremely small, fractals again make their presence felt. Quantum mechanics, the theory governing the behavior of the tiniest particles, is steeped in probabilities and uncertainties. Yet, some propose that beneath this apparent randomness, underlying patterns—fractal in nature—might govern quantum interactions.

Cosmic Implications: A Unified Theory?

The presence of fractals, from the largest cosmic scales to the tiniest quantum realms, suggests an inherent interconnectedness in the universe. This has led some scientists and philosophers to speculate about a potential

Unified Field Theory—a single framework that can explain all physical phenomena, from the motion of galaxies to the dance of subatomic particles. If fractal patterns govern both these extremes, they might be the key to unlocking this unified understanding.

Philosophical and Spiritual Ramifications

Beyond the realms of hard science, the fractal nature of the cosmos touches philosophical and spiritual dimensions. The idea that every part of the universe, no matter how small, reflects the whole resonates with many spiritual teachings that emphasize oneness and interconnectedness. If the universe is truly fractal in nature, then every star, planet, and being holds within it a reflection of the entire cosmos.

Conclusion

Fractals, in their intricate beauty and repeating patterns, offer a tantalizing glimpse into the grand design of the cosmos. Their presence, from the vast cosmic web to the quantum depths, suggests a universe that, despite its apparent chaos, follows an order, a pattern, a design. In this design, every part is intrinsically connected to the whole, echoing the patterns of existence across scales and realms.

As we continue our quest to understand the universe, fractals stand as a testament to the cosmos's inherent beauty and order. They challenge us to look deeper, to find patterns in chaos, and to recognize the interconnected dance of existence that plays out from the smallest quantum particle to the vast, spiraling galaxies. In the grand design of the cosmos, fractals might just be the universal language, telling the timeless tale of unity, order, and infinite wonder.

Chapter 4

Dimensions and Scales: The Infinite Layers

Navigating the Fractal Scales: From Quarks to Quasars

THE UNIVERSE IS A vast and intricate tapestry woven with patterns that repeat across scales, from the tiniest particles known to physics to the most colossal cosmic structures. The idea of navigating through these scales offers a stunning perspective on nature's unity and complexity. As we zoom in from the scale of quasars down to quarks, we venture on a journey that crosses the boundaries of physics, inviting us to explore the fractal-like nature that seemingly connects different layers of the universe.

Quasars: Beacons of the Early Universe

Quasars stand as lighthouses in the cosmos. They are the intensely luminous cores of distant galaxies powered by supermassive black holes. As matter spirals into these black holes, it heats up. It emits powerful jets of radiation that can be seen across great distances, providing clues about the formation and growth of galaxies in the early universe. These celestial objects offer insights into the gravitational forces at play and the behavior of matter under extreme conditions.

Galactic Structures: The Cosmic Web

Zooming in closer, the structure of galaxies themselves displays a fractal-like pattern. Galaxies are not scattered randomly throughout the universe but are connected by dark matter filaments, forming a gigantic cosmic web. This network features voids and nodes where galaxies cluster, creating patterns reminiscent of other complex systems seen in nature, such as neurons in the brain or branching trees, suggesting a scale-invariant structure.

Stellar Nurseries: Clouds of Creation

The birthplaces of stars, known as nebulae, within galaxies are another example of natural fractals—these immense clouds of dust and gas collapse under gravity to form stars. The intricate structures within nebulae, much like the Lichtenberg figures or the veins in a leaf, display a chaotic yet patterned nature that is fundamental to star formation processes. These clouds are not uniform but have a complex topology that can be described using the mathematics of fractals.

The Solar System: Order in Chaos

Descending further, our own solar system is a model of order derived from apparent chaos. The gravitational relationships and orbital dynamics display rhythmic patterns and stability derived from the primordial cloud of gas and dust. The distribution of planetary bodies, their moons, and the asteroid belts hints at underlying resonances and fractal patterns that govern their long-term stability and evolution.

Planetary Geology: Fractal Patterns on Surfaces

The surfaces tell their own tales of fractal patterns on the scale of individual planets. Mountain ranges, river networks, and coastline profiles exhibit self-similarity at different scales. The erosion patterns and the branching networks of rivers and streams mimic the veins of leaves, the branches of trees, and the neural networks of animal brains, suggesting a universal set of rules for shape formation across the natural world.

Life Itself: From Cells to Ecosystems

Life on Earth, from the structure of the cells to the organization of ecosystems, also mirrors fractal geometry. The branching of trees, the structure of blood vessels and bronchi, and the organization of ecological boundaries—all exhibit fractal characteristics. These structures are optimized for maximal interaction with their environment, whether for the exchange of gases in the lungs or the maximization of sunlight captured by leaves.

Molecules and Atoms: The Dance of Particles

On a molecular level, the arrangement of atoms within molecules and their interaction through chemical bonds demonstrate a fractal essence in how electrons are distributed and shared. Molecular structures, especially large complex proteins and DNA, show patterns that are self-replicating through biological processes—patterns that are essential to life.

Subatomic Worlds: Quarks and Quantum Foam

At the smallest scales known to current physics, subatomic particles such as quarks exhibit behaviors that challenge our classical understanding of physics. Quarks combine to form protons and neutrons and are never found in isolation. This confinement is dictated by the strong nuclear force, which is itself defined by complex rules that determine how matter behaves at these infinitesimal scales.

Connecting the Scales

The journey from quarks to quasars is a reminder of the interconnectedness of all things. The patterns seen at one scale often find echoes at others, suggesting that the laws governing the cosmos have a fractal nature. This fractal nature may not be literal in every aspect but serves as a powerful metaphor for understanding the complexity and beauty of the universe.

Understanding the universe through its fractal scales, from quarks to quasars, allows us to appreciate the remarkable balance between randomness and order, chaos and structure, that defines our cosmos. It highlights the deep symmetries and the simple rules that can generate infinite complexities, providing a profound perspective on the universe and our place within it.

The Role of Dimensions in Understanding Fractality

Fractality is a concept that captures the complex, self-similar patterns observable in nature and mathematical models, spanning from the coastline of a continent to the branches of a tree. Understanding fractality through the lens of dimensions offers a profound insight into how these patterns form, function, and facilitate our understanding of complex systems. The dimensionality of a fractal is one of its most intriguing aspects, often being not an integer but a fractional number; hence, the term "fractal" is derived from the Latin word "fractus," meaning broken or fractured.

Defining Dimensions in Fractal Geometry

In traditional Euclidean geometry, dimensions are defined as integer values: a line is one-dimensional, a plane is two-dimensional, and a volume is three-dimensional. However, fractals do not adhere to these neat categorizations. They are characterized by self-similarity across scales: magnify a part of a fractal, and you find similar structures as the whole. This property leads to dimensions that are not whole numbers but fractions, indicating how completely a fractal fills the space.

The concept of fractional dimensions was first introduced by Lewis Fry Richardson and later formalized by Benoît Mandelbrot, who asked how long the coastline of Britain was. Richardson had observed that the measured length of a coastline would increase as the unit of measurement decreased due to the intricate, repeating patterns that become visible at smaller scales. Mandelbrot used this observation to introduce the idea that such irregular shapes could have dimensions that lie between integers.

Fractal Dimensions and Their Calculation

The fractal dimension, or Hausdorff dimension, quantifies the complexity of a fractal—how it scales differently from Euclidean shapes as one zooms in on it. One common method to calculate the fractal dimension is the "box-counting" method, where the space occupied by the fractal is covered with boxes of a certain size, and the number of boxes needed to cover the fractal is counted. As the size of the boxes decreases, the relationship between the number of boxes and the size of each box provides an estimation of the fractal dimension.

This dimension gives insight into how much space a fractal occupies in the embedding space. For instance, whereas a line has a dimension of 1, a perfect plane-filling curve like the Hilbert curve would approach a dimension of 2, filling the plane more completely than a regular curve but less completely than a solid shape.

Implications of Fractal Dimensions

Understanding the fractal dimensions of natural and human-made systems has practical and theoretical implications across various fields:

- **Ecology and Geography:** The fractal nature of river networks, vegetation patterns, and landscape formations helps model ecological interactions and predict geographical formations. For example, the fractal dimension of a river system can provide insights into its overall health and ability to support diverse ecosystems.

- **Medicine and Physiology:** The human body exhibits fractality, from the structure of the lungs and vascular systems to neural dendritic patterns. Understanding these dimensions helps create more accurate models in biomedicine and improve diagnostic technologies.

- **Material Science:** Fractal dimensions are used to describe the porosity and structural behavior of materials at the nanoscale. This has implications for developing materials with specific mechanical properties and for enhancing their strength and durability.

- **Physics:** In theoretical physics, fractal dimensions enter into the scaling laws of turbulence in fluid dynamics and critical phenomena in statistical mechanics. They help describe how systems change at critical points, such as the phase transitions from solid to liquid.

Fractal Dimensions in Art and Technology

In art, fractal dimensions reveal patterns and structures that are aesthetically pleasing and often unexpected. Many artists, such as Jackson Pollock, have intuitively used fractal dimensions to create visually engaging works that keep the viewer's eye moving through the artwork.

In technology, fractal antennas—using the concept of fractal dimensions—allow for efficient and compact design structures used in cellular phones and television receivers. These antennas use self-similar designs to maximize the effective length or increase the perimeter of material that can receive signals within a very small given area.

Conclusion

The study of fractal dimensions bridges the gap between abstract mathematical concepts and practical applications in physical and social sciences. It illuminates how simple, repetitive processes govern natural phenomena that initially seem impossibly complex.

By exploring the role of dimensions in fractality, we gain a deeper appreciation for the inherent beauty and underlying order in the chaos of the natural world, seeing new connections and applications that cross disciplinary boundaries. This exploration is not just about understanding what is but about imagining what could be as we apply these principles to innovate and solve complex problems.

Unifying Space-Time Through Fractal Geometry

The concept of unifying the fabric of space-time through fractal geometry offers a compelling perspective that could bridge the distinct realms of general relativity and quantum mechanics. This convergence could potentially illuminate our understanding of the universe's deepest secrets, from the minuscule quarks to the majestic quasars, through a universal language encoded in fractals.

Fractal Geometry: A Brief Overview

Fractal geometry differs fundamentally from classical geometrical approaches by describing shapes that exhibit self-similarity across different scales and are defined by fractional dimensions rather than integer dimensions. Introduced by Benoît Mandelbrot in the 1970s, fractals have been recognized for their prevalence in natural forms and processes, from the branching of trees and the distribution of galaxies to the formation of coastlines and clouds.

Fractals in Nature and Cosmology

Fractals are not merely mathematical curiosities but are evident in the structure of the universe. The distribution of matter in the universe at large scales shows a fractal-like distribution, with galaxies clustering together in a manner that resembles the complex patterns of a fractal set. This distribution can be seen in the large-scale structure of the cosmos, where galaxy superclusters form filaments and sheets around great voids, suggesting a scale-invariant structure reminiscent of fractals.

Theoretical Foundations of Fractals in Space-Time

The application of fractal geometry to space-time is rooted in the work of scientists who ponder the very nature of the cosmos. Theoretical physicists have proposed models where space-time itself has a fractal structure at the smallest scales, possibly at the Planck length, where the effects of quantum gravity become significant. These models suggest that the smooth space-time continuum postulated by Einstein's theory of general relativity breaks down at quantum scales, giving way to a more fragmented, fractal-like fabric.

Unifying Quantum Mechanics and General Relativity

One of the greatest challenges in modern physics is to unify quantum mechanics, which governs the subatomic world, with general relativity. This theory describes the macroscopic world, including gravity and the dynamics of large celestial bodies and the universe itself. Quantum mechanics and general relativity currently operate in separate theoretical frameworks that are incompatible in extreme environments, such as the singularities of black holes or the conditions at the beginning of the universe.

Fractal geometry offers a promising bridge between these realms. If space-time is fractal, then the universe's geometry at quantum scales could be described in a way that naturally extends to larger cosmic scales. This approach could provide a common language to describe phenomena across all scales, potentially leading to a unified theory of quantum gravity.

Fractals and the Quantum World

In the quantum realm, particles do not exist at specific points but show probabilities of presence, which could be interpreted through fractal patterns. The path of a particle in quantum mechanics, often described as a wave function, might be envisioned as a fractal curve in space-time. This interpretation could radically alter our understanding of particle dynamics and field interactions, portraying them as manifestations of deeper, self-similar geometrical structures.

Implications for Cosmology and Astrophysics

Adopting a fractal-based view of space-time could also have profound implications for cosmology and astrophysics. It might, for instance, offer explanations for the puzzling acceleration of the universe's expansion, dark matter, and dark energy. These phenomena could be understood as integral parts of a fractal universe where the fractal structure of space-time itself influences the laws of gravity and the dynamics of galaxy formation.

Practical Applications and Technological Innovations

Beyond theoretical implications, applying fractal geometry to space-time could lead to new technologies inspired by the inherent efficiencies of fractal designs. In communications technology, fractal antennas—used for their compact size and broadband properties—could be just the beginning. Fractal concepts could enhance computational algorithms, improve the resolution of imaging techniques, and optimize various engineering structures for stability and strength.

Conclusion

The integration of fractal geometry into the understanding of space-time promises to solve some of the most perplexing puzzles in physics. It offers a more profound appreciation of the universe's interconnectedness. This unified approach could prove revolutionary, providing a deeper, more intuitive grasp of the cosmos, where the patterns of a leaf's veins reflect the sprawling arms of a galaxy. As researchers continue to explore this fascinating frontier, the potential to finally unify the fundamental forces of nature lies on the horizon, potentially reshaping our understanding of everything from the microcosm of quantum particles to the macrocosm of the woven cosmos.

Chapter 5

Nature's Repetitive Patterns

Phyllotaxis, Fibonacci, and the Golden Ratio: Patterns in the Human Form

THE NATURAL WORLD IS replete with patterns and mathematical sequences that provide aesthetic pleasure and serve functional roles in the development and evolution of living organisms. Among these, the concepts of phyllotaxis, the Fibonacci sequence, and the golden ratio stand out for their prevalence and significance across various scales of nature, including the human form. These patterns are not merely coincidental; they reveal biological systems' underlying order and efficiency.

Phyllotaxis: The Arrangement of Leaves and Petals

Phyllotaxis refers to the arrangement of leaves on a plant stem or the pattern of seeds in a sunflower. This arrangement is not random but follows specific spiral patterns that maximize sunlight exposure or optimize space. Most commonly, these spirals adhere to the Fibonacci sequence, where each number is the sum of the two preceding ones (1, 1, 2, 3, 5, 8, 13,...). In plants, this sequence ensures that each leaf gets maximum access to sunlight without shading the others, an optimal solution that has evolved through natural selection.

The Fibonacci Sequence and the Golden Ratio

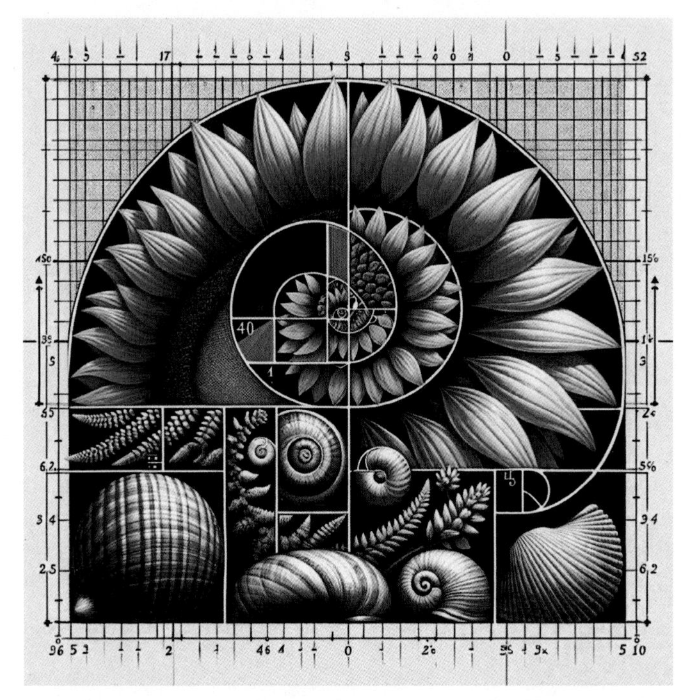

Fibonacci Sequence and the Golden Ratio – 4biddenknowledge Inc

The Fibonacci sequence leads naturally to the golden ratio, denoted by the Greek letter φ (phi), approximately equal to 1.618. This ratio occurs when the ratio of the sum of two quantities to the larger one is the same as the ratio of the larger one to the smaller. The golden ratio is aesthetically pleasing and has been used in art and architecture for centuries, but its occurrence in nature is even more fascinating.

In phyllotaxis, the ratio of successive Fibonacci numbers approximates the golden ratio, and this convergence is often visible in the spiral arrangements of seeds, leaves, and even the branching of trees. These patterns ensure efficient packing and optimal growth, leveraging the mathematical properties of the golden ratio for survival and functionality.

The Human Form and the Golden Ratio

The golden ratio and Fibonacci sequence manifest in various ways in the human body, suggesting a deep biological harmony. Leonardo da Vinci's Vitruvian Man famously encapsulates this idea, illustrating the proportions of the human body and its adherence to the golden ratio.

- **Facial Proportions:** The human face exhibits the golden ratio in numerous ways. The length of the face divided by the width of the face, the distance between the lips and where the eyebrows meet divided by the length of the nose, and several other facial measurements often approximate the golden ratio, contributing to what is perceived as balanced and harmonious facial features.

- **Body Proportions:** The human body itself is a testament to these patterns. The ratio of a person's total height to the height from the ground to their navel often approximates the golden ratio. Similarly, the length of the forearm to the length of the hand and other limb proportions frequently demonstrate this ratio, suggesting an inherent symmetry and balance in human anatomy.

- **DNA and the Golden Spiral:** Even at the microscopic level, the structure of DNA—the molecule that contains the genetic instructions for life—reflects the Fibonacci sequence and the golden ratio. The dimensions of the DNA double helix spiral adhere to the ratio of 21:34, which are consecutive Fibonacci numbers, with the width of the spiral approximating the golden ratio when compared to its length per full cycle of the double helix.

- **Fingerprints and Whorls:** The intricate patterns of fingerprints often show spiraling patterns that can be analyzed through Fibonacci sequences. The way these ridges form and swirl on the tips of our fingers follows the dynamics of differential growth rates, similar to the patterns observed in phyllotaxis in plants.

Implications of These Patterns

The presence of the golden ratio and Fibonacci sequence in the human form is more than an aesthetic matter; it signifies the efficiency of nature's design. This efficiency appears in:

- **Ergonomics and Biomechanics:** Understanding these patterns can enhance ergonomic designs, ensuring that tools and environments fit the natural proportions of the human body more comfortably and effectively.

- **Health and Diagnosis:** In medical imaging and diagnostics, recognizing deviations from typical fractal and golden ratio patterns can help identify pathologies or developmental anomalies.

- **Psychology of Perception:** The innate preference for the golden ratio in human faces and forms plays a role in social perception and aesthetics. Studies have shown that faces and bodies closer to these mathematical proportions are often perceived as more attractive.

- **Art and Design:** Artists and designers have long used the golden ratio to create works that are naturally pleasing to the eye. This understanding helps craft art, architecture, and even everyday objects that resonate with the innate human preference for symmetry and proportion.

Conclusion

The exploration of phyllotaxis, the Fibonacci sequence, and the golden ratio in the human form bridges the worlds of mathematics, biology, and aesthetics. It reveals how deeply interconnected life is, with mathematical principles manifesting in our very anatomy, ensuring functionality, efficiency, and beauty. These patterns underscore a universal language of symmetry and harmony that pervades all levels of existence, from the smallest genetic structures to the proportions of the human body, reflecting a profound unity in the diversity of nature's designs.

Growth, Decay, and the Patterns of Life: The Role of Fractals

Fractals are intricate patterns that repeat at various scales, often found in nature, where they play a crucial role in the processes of growth and decay. These self-similar patterns offer a unique lens to understand life's dynamic and interconnected aspects. From the unfurling of ferns to the erosion of coastlines, fractals provide a mathematical framework to describe how nature evolves, thrives, and eventually declines.

Fractals and the Dynamics of Growth

Growth in nature is rarely linear or uniform; it often follows complex, recursive patterns that can be best understood through fractal geometry. This is evident in the biological world's macroscopic and microscopic realms.

- **Plant Growth:** One of the most striking examples of fractals in growth is seen in plants. The way leaves are arranged, the branching of trees and shrubs, and the structure of root systems often exhibit fractal characteristics. For instance, the branching pattern of trees follows a fractal model to maximize sunlight exposure and nutrient absorption. Each branch, and the branch of that branch, is similar in structure but different in size, optimizing the tree's interaction with its environment.

- **Vascular Networks:** The networks of veins in leaves and the vascular systems in animals (including humans) show fractal patterns. These networks are designed to maximize the efficient transport of nutrients and oxygen. The branching patterns ensure that every cell is adequately supplied, demonstrating how fractal geometry contributes to the functionality of living organisms.

- **Coral Reefs and Marine Life:** Coral reefs are another example where fractal patterns are prevalent. The intricate shapes of corals are not just aesthetically pleasing; they are adaptations to maximize surface area for the photosynthetic algae that live within them. Similarly, the shapes of many marine creatures have fractal dimensions that optimize their movement and interaction with water currents.

Fractals in the Process of Decay

Just as fractals play a role in growth, they are also present in the natural processes of decay and degradation.

- **Erosion and Landforms:** The erosion of landscapes, whether through wind, water, or other natural forces, often follows fractal patterns. The formation of river networks, mountain ranges, and coastlines all exhibit fractal geometry, where the processes of decay and transformation over time remain scale-invariant. This means the patterns observed at one scale can often predict those at another, helping geologists and ecologists understand the evolution of landforms over time.

- **Decomposition in Ecosystems:** The decay of organic matter is fundamental to nutrient cycling in ecosystems. The breakdown of leaves, wood, and other organic materials often shows fractal patterns as microorganisms and other decomposers work at multiple scales. This fractal decomposition is essential for returning nutrients to the soil, ensuring the continuation of life cycles in the ecosystem.

- **Disease and Pathology:** In the medical field, fractal patterns are observed in the progression of certain diseases, especially in the structures formed by cancerous cells. The irregular growth patterns of tumors often exhibit fractal characteristics, and understanding these patterns can lead to better diagnostic tools and treatments.

Patterns of Life: Fractals as a Unifying Theme

Beyond the specifics of growth and decay, fractals serve as a unifying theme across the spectrum of life's processes.

- **Reproduction and Evolution:** The way organisms reproduce and evolve over generations often displays fractal-like patterns. For example, the genealogical trees of populations show self-similar structures that can help trace genetic traits and evolutionary pathways. These patterns are not just historical records but dynamic maps that can predict future evolutionary trends.

- **Population Dynamics:** The distribution and movement of populations, whether animals, plants, or humans, often follow fractal patterns. Urban sprawl, animal migration paths, and even the spread of diseases can be better understood through fractal models, which capture the complexity of these phenomena more effectively than linear models.

- **Behavior and Neural Activity:** On a psychological and neurological level, behavior patterns and brain activity exhibit fractal dynamics. The way thoughts and emotions evolve, the fluctuation of brainwaves, and even the structure of neural networks show self-similarity across scales, offering insights into the fundamental workings of the mind.

Conclusion

The interplay of growth, decay, and the patterns of life through the lens of fractals reveals the profound interconnectedness of all natural processes. By applying fractal geometry, scientists and researchers can decode complex systems, predict changes, and even influence outcomes in fields ranging from ecology to medicine. Fractals provide visual and mathematical beauty and embody the deep symmetries and principles that govern the natural world. Understanding these patterns helps us appreciate the delicate balance of life, the inevitability of change, and the cyclical nature of existence, where every end is a precursor to new beginnings, and every pattern holds the promise of discovery.

Entropy and Fractals

Entropy and fractals are distinct concepts in mathematics and physics, but they intersect in interesting ways, particularly when studying complex systems. Here's a look at how entropy correlates with fractals and the implications of this relationship.

Entropy: A Measure of Disorder

Entropy, in thermodynamics, is a measure of the disorder or randomness in a system. In statistical mechanics, it quantifies the number of ways a system can be arranged while maintaining the same energy level, often interpreted as the system's disorder or randomness. High entropy corresponds to high disorder and low predictability.

In information theory, entropy measures the unpredictability or the amount of information in a message or data set. It quantifies the average rate at which a stochastic source of data produces information.

Fractals: Patterns of Self-Similarity

Fractals are mathematical sets or natural phenomena that exhibit a repeating pattern at every scale. They are characterized by self-similarity and often by a non-integer dimension, known as the fractal dimension. This dimension exceeds the topological dimension and provides a way to quantify the complexity of the fractal pattern.

Correlation Between Entropy and Fractals

- **Entropy in Fractal Structures:** In the study of fractals, entropy can describe the complexity and randomness within fractal patterns. For instance, the entropy of a fractal can give insights into the level of disorder or the amount of information needed to describe the fractal structure. High fractal dimensions often correlate with higher complexity and, in some interpretations, higher entropy.

- **Thermodynamic Entropy and Fractals:** In thermodynamics, certain systems exhibit fractal-like behavior in the way energy is distributed or in the arrangement of particles, especially at critical points of phase transitions. The entropy changes in these systems can reflect fractal characteristics. For example, at the critical point of a phase transition, the system's spatial configuration can be described using fractal geometry, and the entropy associated with this configuration can provide insights into the phase behavior.

- **Information Entropy and Fractal Dimension:** In information theory, entropy is used to quantify the amount of uncertainty or information content. For fractals, the fractal dimension can be related to the information entropy, as it describes how detail in the pattern increases with scale. The higher the fractal dimension, the more information is required to describe the fractal at each scale, and this can be mirrored in the increase in entropy.

- **Entropy in Dynamical Systems with Fractal Attractors:** Fractal attractors such as the Lorenz attractor or the Mandelbrot set can emerge in chaotic dynamical systems. These systems have entropy rates (often referred to as Kolmogorov-Sinai entropy) that measure the rate of information production or disorder. The complexity and unpredictability of these fractal attractors are closely tied to their entropy rates.

- **Entropy in Ecological and Geographical Fractals:** In ecology and geography, the fractal nature of coastlines, vegetation patterns, and landscape structures can be analyzed using entropy to understand the complexity and disorder in these systems. For example, the more irregular and fractal the coastline, the higher its entropy, reflecting its complex structure.

Conclusion

While entropy and fractals come from different scientific disciplines, their intersection provides a powerful tool for understanding the complexity, disorder, and information content in various natural and mathematical systems. This relationship helps in analyzing patterns from the microscopic to the cosmic scale, offering a unified view of disorder and structure in the universe.

Fractal Rivers, Trees, and Clouds: Reading the Patterns of Nature

Nature is a masterful artist, crafting intricate patterns that captivate the eye and challenge the intellect. Among these patterns, fractals stand out for their complexity and ubiquity. Fractals are self-similar structures that repeat at various scales, and they are found in an astonishing array of natural phenomena, from the meandering paths of rivers to the branching of trees and the formation of clouds. By studying these fractal patterns, scientists and mathematicians can gain deeper insights into the processes that shape our world.

Fractal Rivers: The Geometry of Meandering Paths

Rivers are one of nature's most dynamic and visually striking fractal patterns. The meandering of a river is not random but follows a fractal geometry that can be described using mathematical models. These meanders form because of the interplay between water flow, sediment transport, and the erosion of riverbanks. As water flows, it tends to erode the outer banks of bends more heavily than the inner banks, forming sinuous paths that resemble a fractal pattern.

The fractal nature of rivers can be observed in their branching patterns, known as river networks or dendritic drainage systems. These networks exhibit a striking resemblance to the veins of a leaf or the branches of a tree. The fractal dimension of a river network gives insight into the efficiency of water and nutrient transport across the landscape. Higher fractal dimensions indicate a more complex and dense network, which can affect everything from the biodiversity of the river ecosystem to the rate at which pollutants are dispersed or diluted.

Moreover, the fractal analysis of rivers helps understand flood dynamics, predict erosion patterns, and even uncover ancient changes in landscapes due to shifts in river paths. It's a tool that not only provides a window into the past but also aids in the planning and management of water resources.

Fractal Trees: The Architecture of Growth and Survival

Trees are another quintessential example of fractals in nature. The way a tree branches from its trunk to its twigs follows a fractal pattern, which maximizes sunlight exposure and nutrient distribution. Each branch and sub-branch is a smaller, self-similar version of the whole, a pattern that can be described using the concept of fractal dimension.

This fractal branching is not merely an aesthetic feature but a critical adaptation for survival. By following a fractal pattern, a tree can ensure that leaves are spaced optimally to avoid shading each other, thereby maximizing photosynthesis. This pattern also helps in the structural integrity of the tree. The fractal structure distributes weight and stress more evenly throughout the tree, enhancing its resilience to wind and gravity.

In the study of tree growth, fractals also provide insights into the health and age of trees, the efficiency of their resource use, and their role in the carbon cycle. Understanding the fractal geometry of trees has applications in forestry, agriculture, and even urban planning, where the goal is to integrate green spaces effectively into city landscapes.

Fractal Clouds: The Dynamics of the Sky

Clouds, with their ethereal and ever-changing forms, are also fractal in nature. The edges and shapes of clouds exhibit a level of complexity best described using fractal mathematics. This fractal structure results from the turbulent processes that form clouds involving the condensation of water vapor in the atmosphere.

The fractal dimension of clouds can tell meteorologists about the type of cloud and the weather conditions that produced it. For instance, with their cauliflower-like appearance, cumulus clouds have a higher fractal dimension than smooth stratus clouds. This information is crucial in weather prediction and climate modeling, as it affects precipitation patterns, storm formation, and even the climate's response to various atmospheric pollutants.

Studying the fractal geometry of clouds helps in understanding how they interact with light and heat, which is essential for accurate climate models. The way clouds scatter sunlight, absorb heat, and influence global temperature patterns can all be analyzed through their fractal structure.

Implications of Fractal Patterns in Nature

The study of fractal patterns in rivers, trees, and clouds extends beyond academic curiosity. It has practical applications in various fields:

- **Environmental Science and Conservation:** By understanding the fractal patterns in rivers and trees, scientists can better predict the impacts of environmental changes and human activities on ecosystems. This knowledge is vital for conservation efforts, helping to preserve biodiversity and maintain ecological balance.

- **Climate Science and Meteorology:** The fractal analysis of clouds contributes to more accurate weather forecasting and deeper insights into climate change. As models become more refined, predictions about weather and climate impacts on human and natural systems become more reliable.

- **Urban Planning and Architecture:** The principles of fractal geometry are increasingly applied in urban planning and architecture to create efficient and aesthetically pleasing spaces. From the layout of parks to the design of buildings, embracing natural fractal patterns can lead to more sustainable and harmonious urban environments.

- **Art and Design:** Fractals have inspired artists and designers for decades. The natural beauty and complexity of fractal patterns are reflected in art, graphic design, and even fashion, providing endless inspiration and innovation.

Conclusion

Fractals in rivers, trees, and clouds are not just patterns to be admired but are key to understanding the processes that shape our world. These patterns reveal the underlying order in apparent chaos and the interconnectedness of all elements of the natural world. By studying these fractals, we gain insights into the dynamics of growth, decay, and the very essence of life itself. This exploration deepens our appreciation of nature's beauty and enhances our ability to interact with and protect the environment.

As we continue to decode the fractal patterns of nature, we unlock new possibilities for science, technology, and art, bridging the gap between understanding and application in our quest to harmonize with the world around us.

Fractals and Fibonacci in Economic Systems and the Stock Market

The intricate dance of prices in economic systems and the stock market often reveals patterns that are not only predictive but also remarkably consistent with mathematical concepts like fractals and the Fibonacci sequence. These mathematical principles offer a lens through which to understand the complex dynamics of markets, helping traders and economists predict trends and make informed decisions. Here, we explore how fractals and the Fibonacci sequence manifest in economic systems and the stock market, providing insights into their behavior and influencing investment strategies.

Fractals in Economics and the Stock Market

Fractals are self-similar patterns that repeat at various scales, and they have been found to be a fitting description for the behavior of price movements in financial markets. This concept, introduced by Benoît Mandelbrot, posits that financial markets exhibit fractal properties due to the clustering of volatility and the unpredictability inherent in market behavior.

- **Market Volatility and Fractal Patterns:** In financial markets, the prices of assets do not follow a linear path but exhibit jumps and drops in a manner that suggests a fractal structure. This is observed in the way market volatility clusters—large changes tend to be followed by large changes (of either direction), and small changes follow small changes. This clustering gives the market a fractal dimension, where patterns at one-time scale can be indicative of patterns at another.

- **Predictive Power of Fractals:** The fractal nature of the market allows analysts to use past price behavior to forecast future trends. For instance, fractal patterns can be identified and analyzed to determine potential turning points in the market. Techniques like the fractal dimension index and other statistical measures are used to quantify the roughness and complexity of price movements, offering insights into market conditions.

- **Fractals and Market Efficiency:** The fractal view challenges the traditional notion of market efficiency, which assumes that prices reflect all available information. Fractal patterns suggest that markets are far from efficient and exhibit persistent, predictable patterns over time, influenced by collective investor psychology and external shocks.

The Fibonacci Sequence in the Stock Market

The Fibonacci sequence, a series of numbers where each number is the sum of the two preceding ones, appears in various forms in nature and art. In the stock market, Fibonacci ratios, derived from this sequence, are used extensively to predict price movements and define trading strategies.

- **Fibonacci Retracement Levels:** One of the most common tools derived from the Fibonacci sequence is the Fibonacci retracement, which is used to identify potential reversal levels. After a significant price movement, markets often retrace a portion of that move before continuing toward the original trend. These retracement levels, often at 23.6%, 38.2%, 61.8%, and sometimes 78.6% (all derived from Fibonacci ratios), are watched by traders to make "buy or sell" decisions.

- **Fibonacci Extensions and Projections:** Beyond retracements, Fibonacci extensions are used to predict where a price might go following a retracement. These levels can help traders set profit targets or anticipate areas where prices might face resistance or support.

- **Fibonacci Time Zones:** This tool involves dividing the time between two major price peaks or troughs according to Fibonacci ratios to predict when future changes in trend might occur. Although less commonly used than price-based Fibonacci tools, they provide a temporal dimension to Fibonacci analysis.

Real-World Applications and Challenges

- **Behavioral Finance and Fibonacci:** The use of Fibonacci levels in trading reflects principles from behavioral finance, particularly the idea that traders' emotions can lead to predictable outcomes. Fibonacci levels often become self-fulfilling as many traders watch and react to these levels, thereby influencing the market.

- **Fractals and Risk Management:** Understanding the fractal nature of markets can improve risk management. By recognizing the patterns and scales of market movements, traders can set more accurate stop-loss orders and manage their portfolios more effectively, anticipating periods of high volatility and potential large price swings.

- **Algorithmic Trading and Fractals:** In modern markets, algorithmic trading systems often incorporate fractal analysis and Fibonacci levels to execute trades at high speeds. These algorithms can detect patterns faster than human traders, exploiting fractal and Fibonacci insights to gain a competitive edge.

Critiques and Limitations

Despite their utility, applying fractals and Fibonacci in economics and the stock market is not without criticism. Some argue that the retrospective fitting of these patterns to price movements can lead to confirmation bias, where traders see patterns where none exist. Moreover, while these tools can offer insights, they are not foolproof and should be used in conjunction with other analysis methods and a fundamental understanding of market dynamics.

Conclusion

The intertwining of fractals and the Fibonacci sequence with economic systems and the stock market is a testament to the hidden order within apparent market chaos. These mathematical concepts enhance our understanding of market behavior and provide practical tools for traders and analysts. As markets evolve and become increasingly complex, the fusion of mathematical theory, computational power, and human insight will continue to be essential in deciphering the patterns of life that play out on the trading floor. By embracing these concepts, investors and economists can navigate the intricacies of the market with greater confidence and precision, turning the art of trading into a science informed by the timeless patterns of nature.

Chapter 6

Chaos, Complexity, and Order

The Edge of Chaos: Between Randomness and Order

In the intricate dance of the universe, there exists a fascinating realm where the boundaries between randomness and order blur, known as the "edge of chaos." This concept, pivotal in the fields of chaos theory and complex systems, captures a transitional space where systems exhibit properties of both stability and unpredictability. Understanding this liminal zone offers profound insights into how complexity arises in nature, technology, and even human society.

Chaos and Complexity Defined

Before delving into the edge of chaos, it's essential to clarify the concepts of chaos and complexity. Chaos refers to the behavior of deterministic systems that are highly sensitive to initial conditions, often described by the popular metaphor of the "butterfly effect." A tiny change at one point can lead to vast differences in a later state, making long-term prediction practically impossible despite the system's underlying deterministic rules.

Complexity, on the other hand, involves multiple interconnected parts whose interactions give rise to emergent behavior that cannot be predicted merely by analyzing the individual components. Complex systems are characterized by feedback loops, adaptation, and patterns that emerge from the collective behavior of the system's parts.

The Edge of Chaos: A Critical Transition

The edge of chaos is where complex systems can evolve, adapt, and thrive. It is not a static location but a dynamic balance point where the system is neither fully ordered and predictable nor completely disordered and random. At this edge, systems exhibit a mix of stability and flexibility that is thought to be optimal for the emergence of complexity and evolution.

- **Emergence of Life and Biological Evolution:** In biology, the edge of chaos is often cited as the sweet spot for the evolution of life and intelligence. For instance, cellular automata, simple computational models used to simulate biological and other systems, often show the most interesting and complex behaviors when tuned to this edge. It is here that systems can adapt, learn, and evolve, suggesting that life itself might have emerged and evolved in such a dynamic environment.

- **Neural Networks and Brain Function:** The human brain is another example where the edge of chaos plays a crucial role. Neural networks, both biological and artificial, exhibit optimal processing capabilities when they operate at this edge. The brain's ability to balance between rigid order (which can lead to a lack of creativity and adaptability) and total randomness (which can result in incoherent thoughts and actions) is fundamental to its function and intelligence.

- **Economics and Market Dynamics:** In economic systems, the edge of chaos describes markets that are neither wholly efficient and predictable nor entirely volatile and unpredictable. This balance allows for innovation, growth, and the efficient allocation of resources while still being resilient to shocks and stresses. Financial markets often exhibit fractal patterns and nonlinear dynamics that hover around this edge, where new opportunities and risks continuously emerge.

- **Ecology and Environmental Systems:** Ecosystems demonstrate the edge of chaos through their complex interdependencies and feedback loops among species and their environment. These systems maintain resilience and stability while being adaptable enough to withstand and recover from disturbances. The edge of chaos in ecological contexts ensures biodiversity and the sustainable cycling of resources.

Characteristics of Systems at the Edge of Chaos

Systems that operate at the edge of chaos share several defining characteristics:

- **Adaptability:** These systems can respond to changes in their environment in a way that maintains their integrity while allowing for evolution and growth.

- **Robustness and Fragility:** While robust against certain perturbations, these systems can also exhibit fragility under specific conditions, reflecting a dual nature that protects them yet makes them susceptible to transformative shifts.

- **Emergence:** When viewed in isolation, new properties and behaviors emerge at this edge that are not evident in the system's components. This emergent behavior is a hallmark of complex adaptive systems.

- **Feedback Loops:** Positive and negative feedback loops are critical in maintaining the system's position at the edge of chaos, helping it adapt and respond to internal and external stimuli.

Technological Implications and Innovations

In technology, harnessing the edge of chaos has led to advancements in various fields:

- **Artificial Intelligence and Machine Learning:** Algorithms designed to mimic the edge of chaos can improve learning and adaptation in artificial intelligence systems, making them more efficient and capable of dealing with complex, real-world problems.

- **Network Security and Cryptography:** Understanding the dynamics of the edge of chaos helps design secure yet flexible networks to adapt to new threats and vulnerabilities.

Challenges and Future Directions

While the edge of chaos offers a powerful framework for understanding complex systems, navigating this boundary comes with challenges. Precisely identifying where the edge lies for a specific system is difficult, and maintaining a system at this edge requires careful tuning and constant monitoring.

Future research aims to deepen our understanding of how to control and utilize the edge of chaos for enhancing system performance, resilience, and adaptability. By further exploring this boundary between randomness and order, scientists and engineers hope to unlock new paradigms in computing, ecology, economics, and beyond, transforming our approach to solving some of the most pressing challenges of our time.

Conclusion

The edge of chaos is not just a theoretical concept but a fundamental principle that underlies the behavior of complex systems across the universe. From the neural pathways in our brains to the vast networks of galaxies, this delicate balance between order and chaos is key to the emergence and sustainability of life and intelligence. By studying and harnessing this edge, we can gain insights into the nature of complexity and perhaps even the fundamental workings of nature itself.

Complex Systems and Their Inherent Fractality

Complex systems are an integral part of the natural and human-made world, encompassing everything from ecosystems and weather patterns to economies, social networks, and even the human brain. These systems are characterized by numerous interacting components whose collective behavior cannot be easily predicted from the properties of the individual parts. One of the defining features of many complex systems is their inherent fractality—a quality that reveals the deep interconnections and scale-invariant patterns within these systems.

Understanding Complex Systems

A complex system is a network of many components that interact dynamically, often in nonlinear ways, leading to unexpected behavior and emergent properties. These systems are not just complicated; they exhibit characteristics such as adaptability, self-organization, and robustness, differentiating them from merely intricate but predictable systems.

Fractals, with their self-similar patterns across different scales, provide a powerful metaphor and mathematical model for understanding the structure and dynamics of complex systems. The fractal nature of these systems can be observed in the recursive patterns that appear at various scales, offering insights into their functionality and behavior.

Fractality in Natural Systems

- **Ecosystems and Biodiversity:** Ecosystems are quintessential complex systems featuring a multitude of species interacting with each other and their environment. The structure of these ecosystems often exhibits fractal patterns. For instance, the branching patterns of trees, the distribution of river networks, and the organization of coral reefs all show fractal characteristics. These patterns ensure efficient nutrient cycling, energy flow, and resilience against disturbances, illustrating how fractality contributes to the sustainability of ecosystems.

- **Meteorological Systems:** Weather systems are inherently complex and dynamic, with fractal patterns manifesting in cloud formations, wind patterns, and precipitation. The outlines of clouds, the structure of snowflakes, and the distribution of rainfall are examples of fractal geometry in action. These fractal patterns help meteorologists understand and predict weather phenomena, including the formation of storms and the dynamics of climate systems.

- **Geological Formations:** The Earth's surface is shaped by a range of processes, from tectonic movements to erosion, many of which produce fractal patterns. Mountain ranges, coastlines, and river basins exhibit fractal geometry, influencing everything from the distribution of habitats to the flow of rivers. Understanding these patterns helps geologists and environmental scientists map, conserve, and assess hazards.

Fractality in Human-Made Systems

- **Urban Planning and Architecture:** Cities are complex systems of buildings, roads, and human activities, often displaying fractal structures. The layout of streets, the arrangement of neighborhoods, and even architectural designs show self-similar patterns that can influence everything from traffic flow to social interactions. Urban planners and architects use fractal principles to create more efficient and livable spaces.

- **Economic and Financial Markets:** Financial markets are classic examples of complex systems where the actions of millions of individuals result in emergent patterns and trends. Stock market fluctuations, price distributions, and trading patterns often exhibit fractal characteristics, helping economists and traders understand market dynamics and develop investment and risk management strategies.

- **Social Networks and Information Systems:** The digital age has brought about complex systems in the form of social networks and large-scale data structures. The connectivity of individuals in social networks or linking information on the internet often follows fractal patterns, influencing how information spreads and how communities form and interact.

The Role of Fractals in Understanding Complexity

- **Predictive Power:** While complex systems are notoriously difficult to predict due to their sensitivity to initial conditions and nonlinear interactions, fractals provide a way to understand the scaling laws that govern them. By studying the fractal dimensions of various structures within a complex system, scientists and analysts can make more informed predictions about its behavior.

- **Design and Optimization:** In engineering and design, understanding the fractal nature of complex systems can lead to more efficient and robust solutions. For example, fractal-based designs in antennas, materials, and even software algorithms can optimize performance by mimicking the scale-invariant properties of natural fractals.

- **Resilience and Adaptability:** The fractal structure of many complex systems contributes to their resilience. By distributing functionality across different scales and ensuring redundancy, these systems can withstand shocks and adapt to changes without losing their overall integrity. This is crucial for ecosystems, economic systems, and technological networks alike.

- **Interdisciplinary Insights:** The fractal nature of complex systems provides a common language across disciplines, from physics and biology to economics and sociology. This interdisciplinary approach fosters collaboration and innovation, leading to deeper insights into the nature of complexity and the development of new theories and applications.

Challenges and Future Directions

While the study of fractals in complex systems has opened up new avenues of understanding and application, it also presents significant challenges. The inherent unpredictability of complex systems, even with fractal analysis, means there are limits to how much can be predicted or controlled. Moreover, the computational resources required to model these systems in detail are immense.

Future research in this area is likely to focus on enhancing computational models, refining fractal analysis techniques, and exploring the implications of fractality in emerging fields such as artificial intelligence and quantum computing. By continuing to explore the fractal nature of complex systems, scientists and researchers aim to unlock further secrets of the universe, from the microscopic to the cosmic scale, enhancing our ability to navigate and harness the patterns of nature for the benefit of all.

How Nature Optimizes Using Fractal Designs

Nature's efficiency and adaptability are often attributed to evolutionary processes that have honed structures and behaviors over millennia. Among these structures, fractals stand out for their intricate patterns that repeat across scales. Fractal designs in nature are not just aesthetically pleasing; they serve functional purposes, optimizing various biological and ecological processes. This exploration into how nature utilizes fractal designs will reveal the underlying principles that make these patterns effective in natural systems.

Understanding Fractals in Nature

Fractals are patterns characterized by self-similarity, meaning they look similar at any scale. These patterns can be exact, where each smaller unit is a replica of the whole, or statistical, where the overall pattern maintains similarity across scales. Fractals have a non-integer dimension, which describes how a pattern fills space. This unique property allows fractals to occupy space more efficiently than simple geometric shapes, providing solutions to various natural challenges.

Optimization of Light Absorption in Plants

One of the most critical uses of fractal design in nature is in the optimization of light absorption in plants. The fractal structure of leaves and branches allows plants to maximize sunlight exposure. For instance, the branching patterns of trees, where each branch and subsequent smaller branches follow a similar pattern, ensure that leaves are spread out to capture sunlight without excessively shading each other. This fractal distribution is evident in the Fibonacci spirals seen in sunflowers or pine cones, where seeds are arranged in spiral patterns that maximize space usage, ensuring each seed gets adequate light and resources.

Efficient Nutrient Distribution and Transport

Fractals are also key to the efficient distribution and transport of nutrients in organisms. The vascular systems of animals and the xylem and phloem in plants exhibit fractal patterns. In humans, the branching of blood vessels and bronchial tubes optimizes blood and airflow, ensuring that these essential resources reach every cell efficiently. The fractal design minimizes resistance and maximizes surface area within a constrained volume. This principle is critical in the lungs, where a large surface area is needed for gas exchange in a relatively small volume.

Enhanced Strength and Resilience

Fractal designs contribute to the structural strength and resilience of natural forms. The fractal geometry of tree branches and roots helps distribute mechanical stress, allowing trees to withstand high winds and support substantial weight. Similarly, the fractal structure of mountain ranges and river networks contributes to the stability of landscapes, dispersing energy and materials in a way that maintains the integrity of the ecosystem even in the face of environmental changes.

Surface Area Maximization for Heat Regulation

Many organisms use fractal designs to maximize surface area for effective heat regulation. For example, mammalian intestines' convoluted, fractal-like structure maximizes the surface area for nutrient absorption. In colder climates, animals like polar bears have fractal fur patterns that trap heat efficiently. Similarly, the intricate patterns of coral reefs create a large surface area that facilitates the exchange of gases and nutrients, supporting a diverse range of marine life.

Camouflage and Predation

Fractals play a role in camouflage and predation strategies. For instance, the irregular, fractal edges of leaves blend seamlessly with similar patterns in their environment, providing cover for insects and other small animals. Predators, like tigers, have fractal stripe patterns that disrupt their outline, making them less visible in the dappled light of their natural habitat. Using fractals in camouflage demonstrates how these patterns can optimize survival strategies by blending or disrupting visual perception.

Hydrodynamics and Aerodynamics Optimization

In aquatic and aerial environments, fractal designs optimize hydrodynamics and aerodynamics. The fractal edges of some fish scales create vortices that reduce drag, allowing for swift, efficient movement through water. Birds' feathers have a fractal structure that optimizes airflow, enhancing lift and maneuverability. These adaptations show how fractal geometry can optimize movement and energy expenditure in different mediums.

Facilitating Pollination and Reproduction

In the realm of pollination and reproduction, fractals ensure the efficient transfer of genetic material. The intricate fractal patterns of flowers are designed to attract pollinators, guiding them to the nectar and pollen. These patterns often contain ultraviolet markers invisible to the human eye but clear to insects, optimizing the process of pollination. Similarly, the fractal arrangement of spores or seeds enhances their distribution by wind or water, ensuring the spread of the species.

Implications for Technology and Human Design

Understanding how nature uses fractal designs to solve complex problems has significant implications for human technology and design. Biomimicry, the practice of taking inspiration from nature to solve human challenges, has led to innovations such as fractal-based antennas, which use less space and provide better signal reception than traditional designs. In architecture, fractal patterns have been used to create structurally sound and aesthetically pleasing buildings, optimizing the use of materials and space.

Challenges and Future Directions

While the benefits of fractal designs in nature are clear, translating these principles into human technology and systems presents challenges. Replicating the complexity and adaptability of natural fractals requires sophisticated modeling and materials. Additionally, understanding the full scope of fractals in biological systems often requires interdisciplinary research spanning biology, physics, chemistry, and computer science.

Future research will likely focus on uncovering more ways in which fractals contribute to the functionality of natural systems and how these principles can be applied in sustainable technology, medicine, and urban planning. By continuing to explore and emulate the fractal patterns of nature, humans can develop efficient and sustainable solutions, reflecting the elegance and ingenuity of the natural world.

Conclusion

Fractals are a fundamental aspect of nature's toolkit for optimizing the functionality of systems across all scales. From the arrangement of leaves on a plant to the structure of entire ecosystems, fractals ensure that natural systems operate with efficiency, resilience, and adaptability. Understanding and applying these principles offers a pathway to innovative solutions that harmonize with the natural world, enhancing our ability to navigate and sustain our environment.

Chapter 7

Consciousness and the Holographic Brain

THE HUMAN BRAIN RESIDES in complete darkness, isolated and unaware of the world outside. It relies on its trusted allies, collectively known as our sensory perception system, to bridge this gap. The brain dispatches these friends to gather vital information about its surroundings. Sight, hearing, touch, smell, and taste spring into action, each gathering data from the external environment. However, these friends are not equipped to interpret the data themselves; their role is solely to collect and transmit it. Once received, the brain meticulously deciphers this data and constructs a hologram from it. Guided by this holographic projection, we navigate through the third dimension, interpreting our world based on the information provided by our sensory perception system.

The Neural Network: A Fractal Perspective

Neural biological and artificial networks represent a fascinating intersection of biology, technology, and mathematics. Traditionally studied through the lens of neurobiology or computer science, viewing neural networks from a fractal perspective offers profound insights into their structure, functionality, and potential. Fractals, with their self-similar patterns and intricate scaling properties, provide a unique framework for understanding the complexities and capabilities of neural networks.

Fractality in Biological Neural Networks

Biological neural networks, such as those in the human brain, exhibit fractal characteristics at various scales, from the arrangement of neurons to the architecture of neural pathways. This fractality is not merely aesthetic but has functional implications for how the brain processes information and adapts to new challenges.

- **Structure and Organization:** The fractal nature of neural networks in the brain is evident in the branching patterns of neurons and the dendritic trees that connect them. These structures ensure efficient communication and signal processing across vast numbers of neurons. The fractal dimension of these networks influences how signals are propagated and integrated, optimizing the brain's computational capacity.

- **Scaling and Efficiency:** The fractal organization of neural networks allows the brain to maximize its processing power within the limited physical space of the skull. This fractal scaling ensures that neural pathways can handle a wide range of stimuli and tasks without significant loss in efficiency. It also contributes to the brain's remarkable ability to store vast amounts of information in a highly compressed format.

- **Adaptability and Learning:** The fractal geometry of neural networks plays a crucial role in the brain's adaptability and learning processes. By following fractal patterns, neural connections can reorganize dynamically in response to new information or experiences, a process known as neuroplasticity. This fractal reorganization facilitates learning and memory, allowing the brain to evolve its structures in response to its environment.

Holographic Projection

The human brain is incredibly fascinating. It resides in complete darkness and has no direct awareness of the external world beyond the skull. However, the brain relies on its companions—our sensory perception systems: sight, taste, touch, hearing, and smell. These senses act as the brain's allies. The brain asks them, "I need to know what's happening out there." The senses respond, "No problem. We'll gather some data for you." They sense the external world, collect information, and transmit it back to the brain through electromagnetic frequencies via the nervous system.

The brain then processes this data and creates a holographic representation of what it believes is happening outside. We navigate the third dimension based on this holographic projection generated from the data collected and sent to the brain.

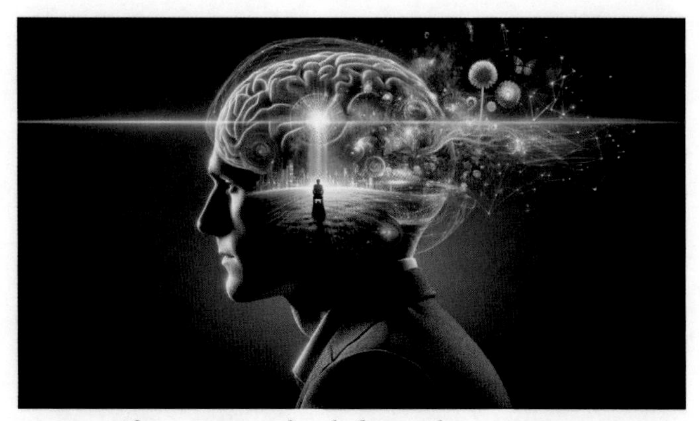

Here is an image of a person with a holographic projection emerging from their mind. The projection represents the external world, highlighting the brain's role in interpreting sensory data. 4biddenknowledge Inc

Fractals in Artificial Neural Networks

Artificial neural networks (ANNs), designed to mimic the functioning of their biological counterparts, also benefit from incorporating fractal principles. These networks are used in a wide range of applications, from image recognition to natural language processing, and understanding their fractal dimensions can enhance their design and performance.

- **Design and Topology:** Incorporating fractal geometry into the design of ANNs can lead to more efficient and robust networks. By mimicking the fractal organization of biological neural networks, ANNs can improve their ability to process complex, high-dimensional data. Fractal-based architectures can enhance the network's capacity to generalize from limited data and adapt to evolving inputs without overfitting.

- **Optimization and Performance:** Using fractal patterns in the training and structure of ANNs helps optimize computational resources and improve performance. Fractal-based methods can reduce the number of connections and parameters needed, making the networks faster and less resource-intensive while still maintaining or even enhancing their accuracy and reliability.

- **Scalability and Robustness:** Fractal structures in ANNs allow for scalable and robust designs that can handle input size and complexity variations. This scalability is crucial for big data and machine learning applications, where the volume and diversity of data continually grow. Networks with fractal architectures are better equipped to manage these challenges, maintaining performance across different scales and conditions.

Implications for Neuroscience and Cognitive Science

The fractal perspective on neural networks has significant implications for neuroscience and cognitive science, particularly in understanding how the brain processes information and generates consciousness.

- **Information Processing:** The fractal organization of neural networks suggests that the brain processes information in a multi-scale, hierarchical manner. This means that understanding cognition and behavior requires looking at how neural patterns at different scales interact and combine to produce emergent properties and functions.

- **Modeling Brain Activity:** Fractal models can help create more accurate brain activity simulations, capturing the complexity of neural interactions more effectively than traditional linear models. These models can assist in studying brain disorders, predicting the effects of neurological interventions, and designing brain-machine interfaces.

- **Consciousness and Cognition:** The fractal nature of neural networks might play a role in the emergence of consciousness and higher-order cognitive functions. By facilitating complex, dynamic interactions within a multi-layered network, fractal structures could be the foundation upon which conscious experience and sophisticated cognitive abilities are built.

Future Directions and Challenges

While the fractal perspective offers exciting possibilities for understanding and designing neural networks, it also presents challenges and open questions for future research.

- **Quantifying Fractality:** Developing methods to accurately measure the fractal dimensions of neural networks, both biological and artificial, is crucial for advancing this field. This quantification can help compare different network architectures and understand the relationship between fractal structure and functionality.

- **Optimizing Fractal Architectures:** For artificial neural networks, finding the optimal fractal architecture for specific tasks remains an area of active research. Balancing the benefits of fractal organization with the computational costs is key to harnessing the power of fractals in practical applications.

- **Understanding Brain Function:** In neuroscience, exploring how fractal geometry influences brain function at various levels—from molecular to behavioral—can provide deeper insights into the workings of the mind. This understanding could lead to breakthroughs in treating neurological disorders and enhancing cognitive abilities.

Conclusion

In conclusion, the intersection of fractals and neural networks opens up a rich landscape of research and application. By embracing the fractal nature of these networks, scientists and engineers can unlock new potentials in understanding the brain and enhancing computational models. As we continue to explore this fascinating convergence, the fractal perspective promises to revolutionize our approach to neural networks, offering a deeper, more integrated view of how complexity and order emerge from the interplay of simple, repeating patterns.

Living in a Holographic Matrix: A Perspective on Reality

The concept that we inhabit a holographic matrix is a fascinating and complex idea that merges science, philosophy, and spirituality. It posits that our reality, as we perceive it, is a construct, much like a hologram, brought to life through the interplay of thought, light, and sound, all operating within an electromagnetic frequency.

Thought and Materialization: The Power of 'I AM'

Central to this concept is the power of thought, encapsulated in the phrase "EYE AM." This notion suggests that our conscious and subconscious thoughts are not merely reflections of our reality but active creators of it. In this framework, thought materializes reality, transforming the abstract—ideas, beliefs, and perceptions – into the concrete.

Reality as a Vibrating Hologram

The theory proposes that what we perceive as solid reality is, in fact, a hologram—a three-dimensional image created by light patterns. These patterns are vibrations brought into existence through electromagnetic frequencies, which are essentially vibrations or sounds at their core. This perspective aligns with the principles of quantum physics, where particles exist in a state of potential until they are observed, collapsing into a quantifiable reality.

The Role of Light and Sound

Light and sound are the fundamental building blocks in this conceptualization of the universe. Everything in our material world is here because our "Higher Light Frequency of Self"—an elevated, more profound aspect of our being—has summoned them into existence. This is analogous to how graphics in a video game materialize as required by the characters on the screen.

Collective Consciousness and Instantaneous Communication

The idea suggests that we are part of a "Collective Consciousness of Everything Created." This collective consciousness encompasses all beings and entities in the universe. It operates on the principle that particles, regardless of their distance from each other, communicate instantaneously. This phenomenon echoes the concept of quantum entanglement, where particles remain connected so that the actions performed on one immediately affect the other, no matter the distance.

Illusion of Separation

A key tenet of this theory is that separation is an illusion. While we may perceive ourselves as distinct from others and the universe, this perspective argues that such separation is false. We are part of a unified whole—a singular entity that cannot be divided. This idea resonates with various spiritual and philosophical teachings emphasizing the interconnectedness of all life and existence.

The Sims Analogy

Comparing our existence to being characters in a simulation game like The Sims provides a metaphor for this concept. Just as Sims are unaware that they are part of a programmed reality, we, too, might be part of a larger design or pattern, functioning within a prearranged set of parameters and rules that we perceive as our reality.

Fractals of Light and the Holographic Universe

The notion that we are part of a fractal of light making up the entire third dimension and the Holographic Universe is an extension of this idea. A fractal is a never-ending pattern, infinitely complex and self-similar across different scales. This suggests that each individual is a microcosm of the larger cosmos—a small part containing the whole.

Consciousness and the Collapse of Reality

In this paradigm, our consciousness transforms electromagnetic energy into quantified bits of information, which we interpret as matter. This aligns with the idea of the observer effect in quantum mechanics, where the act of observation affects the system being observed. Our consciousness, therefore, is not just a passive receiver but an active participant in creating the reality we experience.

Conclusion

The theory of living in a holographic matrix offers a unique lens through which to view our universe and our place within it. It suggests that our perceived reality is an intricate interplay of thought, light, sound, and consciousness, challenging the conventional notions of space, time, and material existence. While this perspective remains largely speculative and philosophical, it echoes many principles found in quantum physics and aligns with various spiritual beliefs about the nature of reality and consciousness.

Quantum Brain Theories and the Holographic Paradigm

In the quest to understand human consciousness and the intricate workings of the mind, scientists and philosophers have often ventured into the realms of quantum mechanics and holographic theory. Quantum brain theories and the holographic paradigm represent bold, interdisciplinary approaches that attempt to explain the complexities of consciousness and cognitive processes in terms of quantum mechanics and holography. These theories propose revolutionary ideas about the nature of reality, the structure of the brain, and the mechanisms that underlie human cognition.

Quantum Brain Theories: Bridging Mind and Matter

Quantum brain theories suggest that quantum mechanics plays a crucial role in the functioning of the brain and the emergence of consciousness. Unlike classical physics, which deals with the macroscopic world, quantum mechanics governs the behavior of particles at the smallest scales—atoms and subatomic particles. Proponents of quantum brain theories argue that the brain's neuronal processes involve quantum phenomena, which could explain the seemingly inexplicable aspects of consciousness.

- **Quantum Coherence and Superposition:** At the heart of quantum brain theories is the concept of quantum coherence and superposition. Quantum coherence refers to the ordered behavior of particles that remain in phase, while superposition is a particle's ability to exist simultaneously in multiple states. Theorists like Roger Penrose and Stuart Hameroff have proposed that microtubules within brain cells maintain quantum coherence, allowing for superposition states that could underlie the parallel processing and integrative features of consciousness.

- **Entanglement and Nonlocality:** Another key aspect is quantum entanglement. In this phenomenon, particles become interconnected so that the state of one particle instantly affects the state of another, regardless of the distance separating them. This property of nonlocality is hypothesized to facilitate the rapid, holistic processing of information in the brain, enabling the unified experience of consciousness.

- **Quantum Information Processing:** Quantum brain theories also suggest that the brain may operate like a quantum computer, processing information through quantum bits (qubits) rather than classical bits. This quantum processing could potentially explain the efficiency and speed of cognitive functions, including memory, perception, and decision-making.

The Holographic Paradigm: A New View of the Brain

The holographic paradigm is another intriguing theory that seeks to explain the brain's functioning and the nature of consciousness. Inspired by the properties of holograms, this paradigm posits that the brain operates in a manner akin to a holographic process, with each part containing information about the whole.

- **Memory and Holographic Storage:** One of the most compelling aspects of the holographic paradigm is its explanation of memory storage. Unlike traditional views that assign specific memories to distinct neuronal locations, the holographic theory suggests that memories are distributed across the brain's entire neural network. This means that each part of the brain contains some part of every memory, much like a piece of a hologram, which still contains the whole image even when cut into fragments. This distributed nature of memory could explain the robustness of memory recall and the ability to remember even with damage to parts of the brain.

- **Perception and Holographic Reconstruction:** In the holographic view, perception is not just a passive reception of sensory data but an active reconstruction process akin to the way a hologram is formed. The brain uses interference patterns from waves of neural activity to construct a third-dimensional perception of the world. This approach could explain the richness and depth of our perceptual experiences, as well as phenomena like phantom limb sensations and the filling in of visual blind spots.

- **Consciousness and the Holistic Brain:** The holographic paradigm extends its reach into the nature of consciousness itself, proposing that consciousness emerges from the holographic nature of brain processes. This means that consciousness is not localized but is a pervasive, emergent property of the entire neural hologram. This holistic view aligns with experiences of unity and interconnectedness reported in various altered states of consciousness.

Challenges and Implications

While quantum brain theories and the holographic paradigm offer fascinating insights, they are not without their challenges and criticisms. The primary challenge is the lack of direct empirical evidence linking quantum processes or holographic phenomena to brain function and consciousness. The delicate and transient nature of quantum states raises questions about their stability in the warm, noisy environment of the brain.

- **Technological and Experimental Advances:** Advances in technology, such as quantum sensors and imaging techniques, are needed to detect and measure quantum effects in biological systems. Similarly, new methodologies are required to validate the holographic model of brain function, potentially through advanced neuroimaging and computational modeling.

- **Philosophical and Theoretical Implications:** These theories challenge conventional views of consciousness and cognition, pushing the boundaries of what is understood about the mind-body problem. They suggest that consciousness might be a more fundamental aspect of reality than previously thought, potentially bridging the gap between physical processes and subjective experiences.

- **Practical Applications:** If validated, quantum brain theories and the holographic paradigm could revolutionize approaches to artificial intelligence, neuropsychology, and the treatment of cognitive disorders. They could lead to developing quantum-inspired algorithms and devices that mimic the brain's efficiency and flexibility.

Conclusion

In conclusion, quantum brain theories and the holographic paradigm represent bold attempts to unravel the mysteries of consciousness and cognition. By exploring the intersection of quantum mechanics, holography, and neuroscience, these theories offer a new lens through which to view the complexity of the human mind. While they face significant challenges, their potential to transform our understanding of the brain and consciousness makes them a vital area of ongoing research and debate in the scientific community.

Could Consciousness Be a Holographic Fractal?

The quest to understand consciousness has long perplexed scientists and philosophers alike. As we delve deeper into the mysteries of the mind and the universe, one intriguing hypothesis emerges: could consciousness itself be a holographic fractal? This idea combines the principles of holography and fractal geometry to propose a revolutionary view of consciousness that transcends traditional boundaries between the physical and the mental, the macroscopic and the microscopic.

Understanding Holographic Fractals

To explore this hypothesis, we first need to understand what holographic fractals are. A fractal is a complex pattern that exhibits self-similarity across different scales. This means that the pattern can be divided into parts, each of which is a smaller copy of the whole. Fractals are found throughout nature, in the branching of trees, the structure of snowflakes, and the distribution of galaxies.

A hologram, on the other hand, is a three-dimensional image formed by the interference of light beams from a laser or another coherent light source. One of the key properties of a hologram is that each part of the hologram contains the information for the entire image. This means that even a small fragment of a holographic film can reconstruct the whole image, albeit with less resolution.

Combining these ideas, a holographic fractal would be a structure that combines the self-similarity of fractals with the whole-in-every-part nature of holography. In the context of consciousness, this concept suggests that the mind could operate as a holographic fractal, where each part reflects the whole and patterns repeat across different levels of cognition and perception.

Consciousness as a Holographic Fractal

- **Multi-Level Self-Similarity:** If consciousness is a holographic fractal, it implies that the processes and patterns of thought, perception, and emotion exhibit self-similarity across different levels of mental activity. This could mean that the way we process information on a subconscious level mirrors the processes at the conscious level but on a smaller scale. Such a structure would allow for a seamless integration of information across the conscious and subconscious realms, facilitating a coherent experience of the world.

- **Information Encoding and Retrieval:** The holographic aspect suggests that every part of the brain (or mind) contains information about the whole. This could explain the robustness of memory and the ability to recall detailed experiences from seemingly fragmentary cues. It could also account for the phenomenon of intuition or insight, where a complex understanding emerges from the subconscious mind as if the brain accesses a holistic view from its distributed fragments.

- **Adaptive and Emergent Properties:** A holographic fractal model of consciousness would inherently possess adaptive and emergent properties. As experiences accumulate and new information is integrated, the fractal structure evolves, reflecting changes not just in content but also in the pattern of interconnections. This dynamic adaptation could explain the growth of personal identity and the development of complex cognitive functions over time.

- **Unity of Experience:** The model could also provide a framework for understanding the unity of consciousness—the seamless way in which we experience a range of sensations, thoughts, and emotions as a single, unified field of awareness. The fractal holographic nature would mean that this unity is not just a surface phenomenon but a deep structural feature of how the brain processes information.

Implications for Neuroscience and Cognitive Science

If consciousness is indeed a holographic fractal, this has profound implications for fields ranging from neuroscience to artificial intelligence.

- **Neuroscience Research:** Traditional neuroscience often looks for specific areas of the brain responsible for particular functions. A holographic fractal perspective would shift this approach to look for patterns of activity that repeat across different scales and regions of the brain. This could lead to new insights into brain plasticity and the mechanisms underlying various cognitive functions and disorders.

- **Cognitive Models:** In cognitive science, this perspective could lead to the development of new models for understanding and simulating human cognition. These models would need to account for the multi-level, self-similar processing of information and the distributed yet coherent nature of mental activity.

- **Philosophy of Mind:** Philosophically, the holographic fractal view challenges the dualistic separation of mind and body, suggesting instead that consciousness emerges from the complex interplay of physical processes at multiple scales. This could bridge some gaps between materialist and idealist views of consciousness.

Challenges and Future Directions

While the idea of consciousness as a holographic fractal is intriguing, it also presents significant challenges.

- **Empirical Evidence:** One of the main challenges is the lack of direct empirical evidence for the holographic or fractal nature of consciousness. Advances in neuroimaging and computational modeling may provide the tools needed to test these ideas more rigorously.

- **Complexity of Implementation:** Simulating a holographic fractal system, especially one as complex as human consciousness, is a daunting task. It requires computational resources and algorithms that can accurately mimic the dynamic, adaptive properties of such systems.

- **Integration of Disciplines:** Exploring this hypothesis requires an interdisciplinary approach, combining insights from quantum physics, neurobiology, psychology, and computational science. Building a coherent framework that encompasses these diverse perspectives is a significant challenge.

Conclusion

In conclusion, the concept of consciousness as a holographic fractal offers a radical yet potentially transformative perspective on the nature of the mind. By combining the properties of fractals and holography, this model provides a way to understand the complexity, unity, and adaptability of consciousness. As research in this area progresses, it holds the promise of revealing new dimensions of the mind and potentially revolutionizing our approach to cognitive science and artificial intelligence. The journey to fully understand this perspective is just beginning, but it represents one of the most exciting frontiers in the quest to decode the mysteries of consciousness.

The idea that the brain connects to or operates within eleven dimensions is a complex and intriguing hypothesis deeply rooted in advanced theoretical physics and neuroscience. This concept often draws on research into high-dimensional spaces in string theory and M-theory, where the universe is posited to consist of more than the familiar three spatial dimensions plus one time dimension.

Recent advances in neuroscience and mathematics have led to the suggestion that the human brain itself might operate in high-dimensional spaces. A study by neuroscientists using algebraic topology, a branch of mathematics that studies shapes and spaces, revealed structures within brain networks that function in up to eleven dimensions. These multidimensional spaces are not physical dimensions like length, width, and height but rather abstract, topological dimensions, which describe how neural connections form highly complex structures — essentially, multidimensional geometrical shapes — that facilitate the processing of information.

This multidimensional complexity helps to explain how the brain can seamlessly integrate vast amounts of information from various sensory inputs and cognitive processes in a coherent manner. The high-dimensional spaces allow for a more nuanced and detailed mapping of neural interactions, leading to richer and more flexible patterns of activity.

This could underlie the remarkable capabilities of the human brain, from the depth of our thoughts to the richness of our perceptions and the creativity of our minds. It's a frontier of research that melds the abstract realms of theoretical physics with the tangible intricacies of brain function, opening up new avenues for understanding cognition and consciousness.

Theoretical physicists, including Michio Kaku, explore the fascinating idea that our universe might exist in eleven dimensions, a concept deeply rooted in the realm of string theory and M-theory. String theory, initially developed to describe the fundamental particles and forces of the universe, posits that what we perceive as particles are actually tiny, vibrating "strings" whose modes of vibration determine their properties. This theory initially required the existence of extra spatial dimensions beyond the familiar three to be mathematically consistent.

M-theory, an extension of string theory, takes this concept further by suggesting that these strings could also be higher-dimensional membranes, or "branes," moving through an eleven-dimensional space-time. Michio Kaku, among others, has been a vocal proponent of this view, explaining that these additional dimensions are not observed in everyday life because they are compactified—that is, they are curled up so tightly at such a small scale that they become effectively invisible to our current observational capabilities.

This eleven-dimensional framework allows physicists to unify the various string theories previously thought to be distinct and offers a more comprehensive understanding of the universe's fundamental structure. It holds the potential to resolve long-standing puzzles in physics, including the unification of gravity with the other fundamental forces and the nature of dark matter and dark energy. By contemplating a universe with eleven dimensions, scientists like Kaku invite us to rethink our understanding of reality, opening up new possibilities for discovery and innovation in both theoretical and applied physics.

Hypothesis: The Universe's Eleven Dimensions and the Brain's Fractal Connectivity

Introduction

The universe, according to advanced theories in physics, may consist of eleven dimensions, as posited by string theory and M-theory. Simultaneously, recent advancements in neuroscience suggest that the human brain might operate within a high-dimensional, fractal-like network. This hypothesis explores the potential connection between these two realms, proposing that the human brain's fractal structure could be a microcosmic reflection of the universe's eleven-dimensional structure, enabling a unique interface between human consciousness and the multidimensional fabric of reality.

Hypothesis Statement

The human brain's fractal network operates within an eleven-dimensional space, mirroring the universe's eleven-dimensional structure as proposed by string theory. This fractal connectivity enables the brain to process and interact with information across these dimensions, potentially explaining the depth and richness of human consciousness and perception.

Theoretical Background

- **String Theory and the Universe's Eleven Dimensions:** String theory suggests that particles are not zero-dimensional points but one-dimensional "strings" vibrating in a higher-dimensional space. M-theory extends this concept to eleven dimensions, where these strings and membranes interact within a complex, higher-dimensional framework. This model offers a unified explanation for all fundamental forces and types of matter.

- **The Brain's Fractal Geometry:** Neuroscience has uncovered that the brain's neural network exhibits fractal patterns with self-similar structures at various scales. This fractality is not just a structural feature but also a functional one, enhancing the brain's computational efficiency and capacity for information processing across different levels of organization.

Connecting the Brain and the Universe

- **Dimensional Resonance:** The hypothesis posits that the brain's fractal network is not limited to three-dimensional space but extends into the higher dimensions suggested by string theory. This eleven-dimensional fractal structure allows the brain to resonate with the same dimensional framework that underlies the universe, facilitating a form of dimensional resonance that could underpin aspects of consciousness and cognition.

- **Information Processing Across Dimensions:** By operating within an eleven-dimensional fractal network, the brain could theoretically access and process information across these dimensions, much like a computer tapping into a multidimensional database. This could provide a basis for the intuitive leaps, creative insights, and complex problem-solving abilities that are hallmarks of human cognition.

- **Perception and Reality Construction:** If the brain interfaces with eleven dimensions, this could influence how we perceive reality. The brain might not just passively receive sensory data but actively construct a multidimensional experience of the world, integrating information across the hidden dimensions to produce the coherent experience of reality we observe.

Empirical Considerations and Predictions

- **Neuroimaging and Dimensionality:** Advanced neuroimaging techniques could be developed to detect signs of high-dimensional structures within the brain's neural network. Techniques like algebraic topology could help identify patterns indicative of eleven-dimensional connectivity.

- **Behavioral and Cognitive Tests:** Experiments could be designed to test if human cognition exhibits signs of accessing higher-dimensional information, such as through tasks that require integrating complex patterns of data that would challenge purely three-dimensional processing capabilities.

- **Quantum Effects in Neural Processes:** Quantum biology could provide a mechanism for maintaining coherence in high-dimensional brain operations. Observing quantum coherence in brain processes could support the hypothesis of the brain's capacity to operate across multiple dimensions.

Conclusion

This hypothesis bridges the gap between cosmological theories and neurological structures, suggesting that the human brain's fractal geometry could reflect and interact with the universe's eleven-dimensional structure. By exploring this connection, we might deepen our understanding of consciousness, enhance artificial intelligence models, and even unlock new dimensions of human potential. This exploration could ultimately lead to a revolutionary understanding of the mind's place within the universe.

Chapter 8

Fractal Cosmology

The Universe as a Nested Hierarchy: Galaxies within Clusters within Superclusters

THE COSMOS IS A grand theater of structures, each nested within larger, more complex systems. From the individual stars that form the fabric of galaxies to the immense superclusters that dominate the observable universe, each level of the cosmic structure reveals a pattern of organization and interdependence that can be described as a nested hierarchy. This perspective helps illuminate the vastness and intricacy of the cosmos, showcasing the elegance of its construction from the smallest to the largest scales.

Galaxies: The Building Blocks of the Universe

Galaxies are the fundamental units in the cosmic hierarchy, massive systems consisting of billions of stars, along with gas, dust, and dark matter, all bound together by gravity. These stellar cities come in various shapes and sizes, from spiral galaxies like our Milky Way, with their elegant pinwheel arms, to elliptical galaxies, which appear as more uniform, spherical, or ellipsoidal swarms of stars.

Within galaxies, stars themselves are often grouped into smaller systems, such as star clusters and stellar associations, which can range from a few dozen to thousands of stars. These stars and their planetary systems, including our own, represent the most intimate level of the nested cosmic hierarchy, where the processes of planetary formation and stellar evolution unfold.

Clusters of Galaxies: The Next Level of Cosmic Organization

Scaling up from individual galaxies, we encounter galaxy clusters, the next major level in the cosmic hierarchy. Clusters consist of dozens to thousands of galaxies bound together by gravity, often dominated by a few massive galaxies at their center. These clusters are not static but dynamic entities, where galaxies interact and merge under the influence of gravitational forces, leading to a variety of astrophysical phenomena.

Clusters are often permeated by hot, X-ray-emitting gas and threaded with dark matter, which together make up the majority of the cluster's mass. These components exert a profound influence on the cluster's evolution and the fate of the galaxies within it. The interactions within clusters lead to star formation, galactic cannibalism, and the triggering of active galactic nuclei, showcasing a level of interactivity that influences the life cycle of galaxies.

Superclusters: The Largest Known Structures

Beyond clusters lie superclusters, vast networks of galaxy clusters interconnected by filaments of galaxies and dark matter, spanning hundreds of millions of light-years. These superclusters represent the largest coherent structures in the universe and form the next tier in the cosmic hierarchy.

Superclusters are often identified by their dense cores of rich clusters surrounded by less densely populated regions, extending into the cosmic web's voids and filaments. The Milky Way, for example, resides in the Laniakea Supercluster, a massive structure that contains over one hundred thousand galaxies spread across 500 million light-years.

The dynamics of superclusters are governed by the large-scale flows of galaxies toward regions of higher gravitational potential, known as the Great Attractor in some cases. These flows demonstrate the interconnected nature of cosmic structures, as galaxies and clusters move in response to the underlying gravitational landscape shaped by dark matter and cosmic expansion.

Implications for Cosmology and the Nature of the Universe

The nested hierarchy of galaxies, clusters, and superclusters has profound implications for our understanding of the universe and the fundamental laws that govern it.

- **Cosmological Parameters:** The distribution and dynamics of galaxies within clusters and superclusters provide crucial information for determining cosmological parameters like the Hubble constant, dark matter density, and dark energy's effects on cosmic expansion. By studying these structures, astronomers can refine models of the universe's overall geometry and fate.

- **Evolution of Galaxies:** The interactions within and between clusters and superclusters drive the evolution of galaxies. By observing galaxies in different environments—from isolated galaxies to those in dense clusters—astronomers can discern how collisions, mergers, and gravitational interactions influence galactic evolution.

- **Dark Matter and Dark Energy:** The behavior of galaxies within clusters and the overall structure of superclusters offer key evidence for the existence and properties of dark matter and dark energy. The gravitational effects that bind galaxies in clusters and the large-scale structure patterns in superclusters cannot be explained by visible matter alone, pointing to the dominant influence of these mysterious components.

- **Large-Scale Structure Formation:** The nested hierarchy provides a framework for studying the formation of large-scale structures in the universe. Theories like the cold dark matter model with a cosmological constant (ΛCDM) predict the growth of structures from small initial fluctuations to the intricate web of galaxies, clusters, and superclusters observed today. This model helps explain the universe's hierarchical organization from the earliest moments after the Big Bang to the present.

Future Research and Exploration

As observational technology advances, so too does our ability to probe deeper into the nested hierarchy of the universe. Projects like the James Webb Space Telescope and large-scale surveys such as the Sloan Digital Sky

Survey continue to map the cosmos in unprecedented detail, revealing the intricate connections between galaxies, clusters, and superclusters.

Future research will focus on understanding how these structures formed and evolved over cosmic time, the role of dark matter and dark energy in shaping them, and what they can tell us about the universe's ultimate fate. By exploring the nested hierarchy of cosmic structures, astronomers hope to unlock the secrets of the cosmos, from the smallest scales of star formation to the grandest scales of the cosmic web.

Conclusion

In conclusion, the universe as a nested hierarchy—from galaxies to superclusters—provides a captivating view of the cosmos that highlights the interconnectedness and complexity of all cosmic structures. This perspective not only deepens our understanding of the universe's architecture but also offers profound insights into the fundamental processes that have shaped the cosmos throughout its history.

Cosmic Webs and the Interconnectedness of Matter

The universe is an intricate tapestry woven with galaxies, stars, and the invisible threads of dark matter and energy. This vast network, often referred to as the cosmic web, represents the most grandiose scale of structure in the cosmos, illustrating the deep interconnectedness of all matter. The cosmic web is not merely a collection of galaxies strewn randomly across the void; it is a complex, interconnected system where the gravitational dance of galaxies, gas, and dark matter unfolds over billions of years, revealing fundamental truths about the universe itself.

Understanding the Cosmic Web

The cosmic web is a structure composed of galaxies, galaxy clusters, superclusters, filaments, sheets, and voids. This network spans the observable universe, with galaxies and clusters forming at the intersections of filaments and sheets, while vast voids devoid of luminous matter separate these dense regions.

- **Formation and Evolution:** The cosmic web's formation traces back to the earliest moments after the Big Bang. Tiny quantum fluctuations in the density of matter in the early universe grew under the influence of gravity, forming the seeds of the cosmic web's structure. Over billions of years, these initial perturbations evolved into the large-scale structure we observe today. Dark matter, which interacts primarily through gravity, played a crucial role in this evolution, providing the scaffolding on which ordinary, baryonic matter could accumulate and form galaxies and stars.

- **Filaments and Nodes:** The most distinctive features of the cosmic web are its filaments—long, slender structures that connect clusters and groups of galaxies across vast distances. These filaments are formed primarily from dark matter and are the highways along which gas and galaxies flow toward larger structures, known as nodes, where galaxy clusters reside. The nodes and filaments together create a network that facilitates the exchange of matter and energy across the cosmos.

- **Voids and Sheets:** Between the filaments and nodes lie vast voids—expanses of space with very few galaxies. These voids are the largest coherent structures in the universe and are essential for understanding the cosmic balance of mass and energy. Surrounding the voids are sheets, thin planes of galaxies, and dark matter that form the boundaries between voids and the denser parts of the cosmic web.

Interconnectedness of Matter in the Cosmic Web

The cosmic web illustrates the deep interconnectedness of matter in several profound ways:

- **Gravitational Interactions:** At the heart of the cosmic web's structure are the gravitational interactions that bind galaxies, dark matter, and gas. These interactions not only shape the web but also drive the evolution of its components. Gravity pulls matter from less dense areas (voids) into more dense areas (filaments and nodes), fostering the growth of structures and influencing the motion and evolution of galaxies within the web.

- **Gas Flows and Galactic Formation:** The filaments of the cosmic web are not just pathways for galaxies; they are also conduits for gas. Intergalactic gas flows along these filaments, feeding galaxies and fueling star formation. This process shows how interconnectedness at a cosmic scale directly impacts the life cycle of galaxies, from their birth to their evolution and eventual demise.

- **Cosmic Feedback Loops:** The cosmic web is a site of continuous feedback processes. Supernovae explosions in galaxies and energetic outputs from active galactic nuclei can expel gas from galaxies into the intergalactic medium, enriching it and influencing subsequent generations of star and galaxy formation. These feedback loops highlight the interconnected nature of cosmic events, where the activities within galaxies influence their larger environments and vice versa.

- **Dark Matter and Cosmic Connectivity:** Dark matter, though invisible, is the primary architect of the cosmic web's structure. Its distribution defines the skeleton of the web and dictates the flow of baryonic matter. Understanding the role of dark matter is crucial for deciphering the cosmic web's interconnectedness, as it binds the luminous and dark components of the universe in a single, cohesive structure.

Implications for Cosmology and Astrophysics

The study of the cosmic web has profound implications for cosmology and astrophysics, offering insights into fundamental questions about the universe's nature and workings.

- **Mapping the Large Scale Structure:** Detailed maps of the cosmic web, created using surveys of galaxies and the cosmic microwave background, help cosmologists test theories of cosmic evolution and the nature of dark matter and dark energy.

These maps are crucial for understanding how the universe has expanded and evolved over time.

- **Understanding Galaxy Evolution:** By examining how galaxies are distributed within the cosmic web and how they interact with their environment, astronomers can better understand the processes that drive galaxy evolution. This includes the role of environment in shaping galaxy morphology, star formation rates, and the growth of supermassive black holes.

- **Probing the Early Universe:** The structure of the cosmic web carries imprints from the early universe, providing clues about the initial conditions after the Big Bang. By studying the cosmic web, scientists can infer properties of the early universe, including the nature of primordial fluctuations and the influence of cosmic inflation.

Challenges and Future Directions

While our understanding of the cosmic web has grown significantly, numerous challenges remain:

- **Detecting Dark Matter:** Directly observing dark matter within the cosmic web is a major challenge due to its elusive nature. Future observations and experiments aim to reveal more about dark matter's properties and distribution within the web.

- **Simulating the Cosmic Web:** Advanced computational models are required to simulate the cosmic web accurately. These simulations must account for the complex interplay of gravity, dark matter, baryonic physics, and cosmic expansion to mimic the observed universe.

- **Integrating Multi-Wavelength Data:** To fully understand the cosmic web, data from various astronomical observations across the electromagnetic spectrum must be integrated. This includes radio, optical, infrared, X-ray, and gamma-ray observations, each providing unique insights into different components of the web.

Conclusion

In conclusion, the cosmic web represents a remarkable example of the interconnectedness of matter on the grandest scales. It is a structure that binds galaxies, governs the flow of gas and dark matter, and encapsulates the history and future of the universe. By studying this grand network, we continue to unravel the mysteries of the cosmos, deepening our understanding of the fundamental forces and processes that have shaped our universe from the very beginning.

Chapter 9

Professor Sylvester James Gates, Jr. and the Discovery of Adinkra Codes

Pioneering Work in Supersymmetry and Supergravity

Professor Sylvester James Gates, Jr. and the Discovery of Adinkra Codes: Pioneering Work in Supersymmetry and Supergravity

PROFESSOR SYLVESTER JAMES GATES, JR., a distinguished theoretical physicist, has made profound contributions to the fields of supersymmetry and supergravity, most notably through his discovery of Adinkra codes. These discoveries not only deepen our understanding of the universe at its most fundamental levels but also bridge complex mathematical concepts with the intriguing possibilities of a unified theory of physics.

Professor Sylvester James Gates, Jr. - Theoretical Physicist, University of Maryland - College Park, Photo taken at the University of Maryland by John Consoli

Introduction to Sylvester James Gates, Jr.

Sylvester James Gates, Jr., often referred to as Jim Gates is an American theoretical physicist renowned for his research in supersymmetry, supergravity, and related aspects of string theory. His work has been pivotal in exploring and expanding these fields, particularly through his innovative use of mathematical symbols and structures to elucidate complex physical theories.

Supersymmetry and Supergravity: Foundations of Gates' Work

To appreciate Gates' contributions, one must first understand the basics of supersymmetry and supergravity. Supersymmetry is a theoretical framework that proposes a type of symmetry between bosons (particles that carry forces) and fermions (particles that make up matter). This symmetry suggests that there is a corresponding fermion for every boson and vice versa. This idea has profound implications for unifying the forces of nature and solving some of the puzzles left unanswered by the Standard Model of particle physics, such as the nature of dark matter.

Supergravity extends the concept of supersymmetry into the realm of gravity, incorporating it into a unified framework with the other fundamental forces. It is a field theory that combines general relativity (Einstein's theory of gravity) with supersymmetry, aiming to create a more comprehensive theory that could potentially lead to a full theory of quantum gravity.

The Double Slit Experiment

The double-slit experiment is a classic physics experiment that demonstrates fundamental principles about the nature of light and matter and the wave-particle duality in quantum mechanics. Here's a detailed explanation:

Experimental Setup

1. **Light Source:** A coherent light source, such as a laser, is used to emit a beam of light.

2. **Barrier with Two Slits:** A barrier is placed in the path of the light beam. This barrier has two parallel, closely spaced slits.

3. **Screen:** Behind the barrier, there is a screen to detect the light that passes through the slits.

Classical Expectation

If light were only a particle, we would expect it to behave similarly to bullets being fired through two holes. We would see two distinct regions on the screen where the particles hit, corresponding to the two slits.

Wave Behavior

When the experiment is performed, an interference pattern emerges on the screen—bright and dark bands. This pattern is characteristic of wave behavior, where waves passing through the two slits interfere with each other:

1. **Constructive Interference:** Where the peaks of two waves meet, they add up, creating bright bands.

2. **Destructive Interference:** Where the peak of one wave meets the trough of another, they cancel out, creating dark bands.

Quantum Mechanics and Wave-Particle Duality

The interference pattern still appears when the experiment is conducted with particles such as electrons instead of light, suggesting that particles also exhibit wave-like properties. This is where the concept of wave-particle duality comes into play: particles can exhibit properties of both waves and particles.

Observer Effect

A crucial aspect of the double-slit experiment involves observing which slit the particle goes through:

- **Without Observation:** When not observed, particles like electrons create an interference pattern, indicating wave-like behavior.
- **With Observation:** When a measurement device is placed to determine which slit the particle goes through, the interference pattern disappears. Instead, a pattern corresponding to particle behavior (two bands) appears, indicating particle-like behavior.

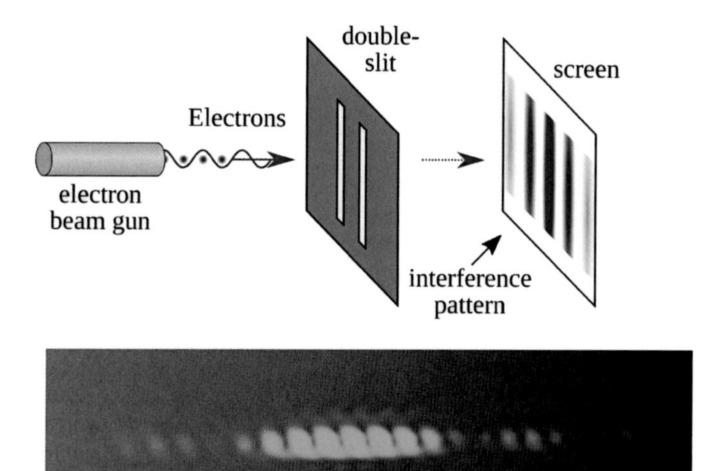

Double Split Experiment - Wikipedia

Implications

1. **Wave-Particle Duality:** The experiment illustrates that light and matter can exhibit both wave and particle characteristics.

2. **Quantum Superposition:** Particles exist in a superposition of states, going through both slits simultaneously until observed.

3. **Collapse of the Wave Function:** Observing which slit the particle goes through collapses its wave function from a superposition to a definite state.

4. **Role of the Observer:** The act of measurement affects the system, a fundamental concept in quantum mechanics.

Conclusion

The double-slit experiment challenges classical intuitions about the nature of reality. It demonstrates that particles can behave as waves and that observation plays a crucial role in determining the state of a quantum system. This experiment is foundational in understanding quantum mechanics and has profound implications for the nature of reality, the role of the observer, and the behavior of particles at the quantum level.

If electrons orbiting atoms can make conscious decisions to convert from a wave into a particle, it suggests that they themselves must possess consciousness. In fact, everything in the universe is conscious, whether it is an animate or inanimate object, man-made or organic because all atoms are conscious. The statement "man-made" is a misnomer. Man doesn't actually make anything in the sense that man creates from scratch. The term “man-made” is a misnomer. Humans do not create anything from scratch; instead, we use existing building blocks called atoms, stacking them like Legos to construct objects such as clothing, buildings, technology, and everything else essential for our civilization. We have learned to build objects using these existing particles. Everything we create possesses some level of consciousness.

Humans building objects – 4biddenknowledge Inc

In this realm, everything exists as a probability or a wave function, a wave of infinite possibilities, until consciousness collapses those superpositions into a single concept, idea, or particle. We are intrinsically linked to creation, as our conscious thoughts help mold and create our reality.

The double-slit experiment provides us with a better understanding of how every frame of existence appears as needed. When we are not at home, our house exists as waves of potential. As we head back and can see it, the house collapses back into the structure we call our home. Even human beings exist as waves of potential and then again as digital bits of matter.

The double-slit experiment has opened our eyes to incredible possibilities and given us a glimpse into the nature of reality itself. We are living in a holographic matrix.

The results of the double slit experiment confirm that wave-particle duality is a fundamental concept in quantum mechanics that describes how every particle or quantum entity, such as an electron or photon, exhibits both wave-like and particle-like properties. This duality is a cornerstone of quantum mechanics, challenging classical physics' straightforward distinction between waves and particles.

The concept of wave-particle duality emerged in the early twentieth century through pivotal experiments and theoretical work:

- **Albert Einstein (1905):** Proposed that light, which was traditionally considered a wave, could also be thought of as consisting of discrete packets of energy called photons. This idea explains the photoelectric effect, where light knocks electrons off a material.

- **Louis de Broglie (1924):** Hypothesized that particles such as electrons could exhibit wave-like properties. This was confirmed by experiments demonstrating electron diffraction, where electrons create interference patterns typical of waves.

Implications for Quantum Mechanics

- **Superposition:** Particles can exist in multiple states simultaneously until observed. The act of observation collapses the wave function into a single state.

- **Uncertainty Principle:** Proposed by Werner Heisenberg, it states that one cannot simultaneously know a particle's exact position and momentum. The more precisely one property is known, the less precisely the other can be known.

- **Quantum Entanglement:** Particles can become entangled, such that the state of one particle instantly influences the state of another, no matter the distance between them.

The holographic principle suggests that our three-dimensional reality is a projection of information encoded on a two-dimensional surface. This idea was first proposed in the context of black hole physics and later extended to the entire universe.

Wave-Particle Duality in a Holographic Universe

- **Information Encoding:** Just as wave-particle duality shows that particles exist as waves of potential until observed, the holographic principle suggests that the universe's information is stored on a lower-dimensional boundary and projected as our perceived reality.

- **Reality as a Projection:** In a holographic model, every point in the three-dimensional space contains information about the entire space, similar to how each part of a hologram contains information about the whole image.

- **Consciousness and Reality Creation:** If particles exist in a superposition of states until observed, it implies that consciousness plays a vital role in shaping reality. Our conscious observation collapses these wave functions, projecting a specific reality from a multitude of possibilities.

Practical and Philosophical Implications

- **Nature of Reality:** The dual nature of particles suggests that reality is not as fixed and solid as it appears. Instead, it is a dynamic interplay of possibilities influenced by observation and measurement.

- **Role of the Observer:** The act of observation is crucial in determining the state of a quantum system. This challenges the classical view of an objective reality independent of the observer.

- **Interconnectedness:** Quantum entanglement and the holographic principle suggest that everything in the universe is interconnected. Changes in one part of the system can instantaneously affect another, regardless of distance.

- **Philosophical Questions:** Wave-particle duality and the holographic principle raise profound questions about the nature of consciousness, the fabric of reality, and the limits of human knowledge.

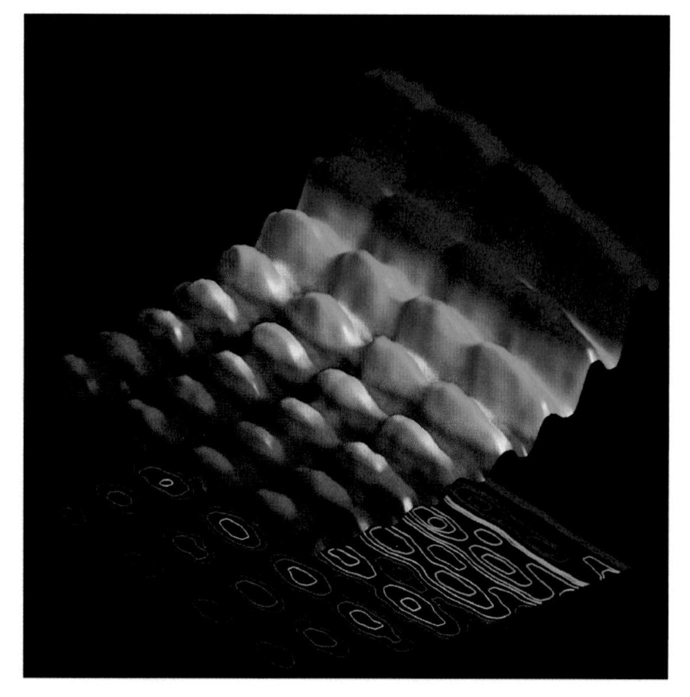

The first ever photograph of light as both a particle and wave – Phys.org https://phys.org/news/2015-03-particle.html

Wave-particle duality reveals the complex and non-intuitive nature of the quantum world, where entities exhibit both wave-like and particle-like properties. When combined with the holographic principle, this duality suggests that our reality may be a projection of information encoded on a lower-dimensional surface. These concepts challenge our traditional understanding of reality, highlighting the significant role of consciousness in shaping the world we perceive and suggesting a deeply interconnected universe.

Discovery of Adinkra Codes

One of Gates and this amazing team's most significant contributions to theoretical physics is their work on Adinkra codes. These are graphical tools used to represent mathematical ideas related to supersymmetry and supergravity. Adinkras are visual representations that help physicists and mathematicians understand complex relationships between different components of supersymmetric theories.

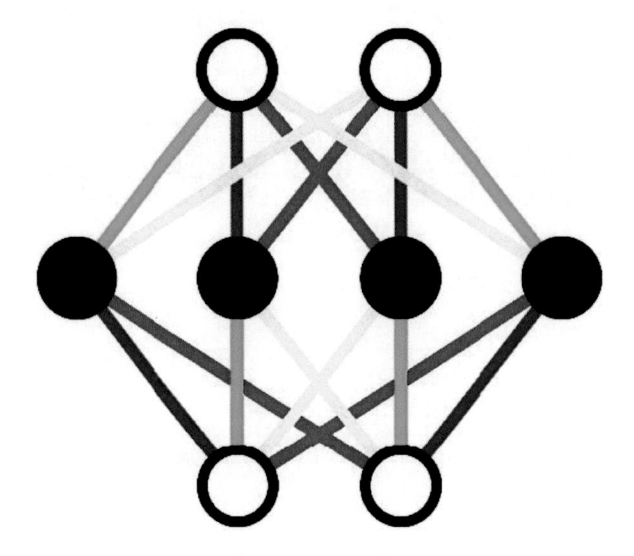

Professor Gates, Michael Faux and Tristan Hubsch plus four mathematicians, Charles Doran, Kevin Iga, Greg Landweber, Robert Miller, worked and collaborated together to unravel the mystery of the Adinkra Codes. This ground breaking discovery was published in the International Press on the 24th of October, 2021. https://www.intlpress.com/site/pub/pages/journals/items/atmp/content/vols/0015/0006/a007/

The term "Adinkra" is derived from symbols used in West African culture, specifically among the Akan people of Ghana. These symbols are rich in philosophical and spiritual significance and are used to convey proverbs and maxims. Gates adopted this term to describe the graphical symbols he developed, which encapsulate the algebraic structures of supersymmetry in a visually intuitive form.

Adinkras are not just aesthetic representations; they are powerful mathematical tools that encode information about the transformation properties of supersymmetric particles. By studying these symbols, researchers can decipher the intricate relationships and symmetries that underpin supersymmetric theories.

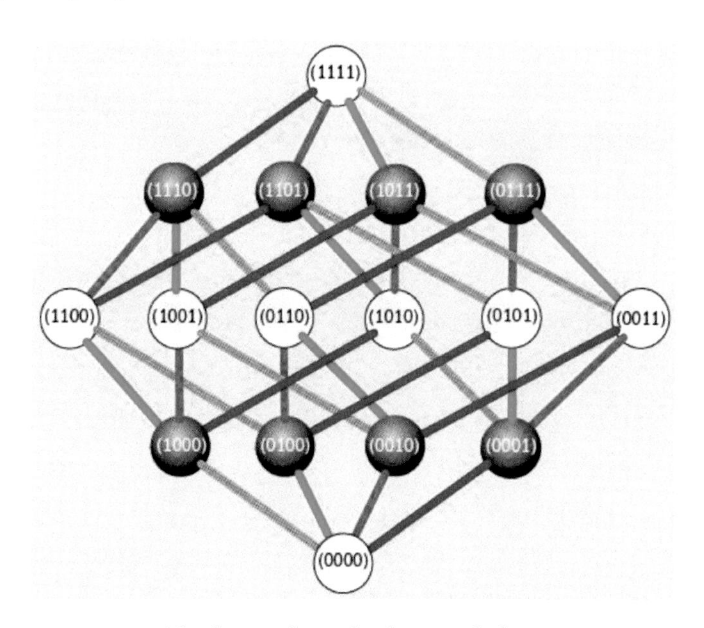

Adinkra code - adinkrasymbols.org

Implications of Gates' Work

Gates' discovery of Adinkra codes has several important implications for theoretical physics and beyond:

- **Clarifying Supersymmetric Structures:** Adinkras provide a clearer, more accessible way to understand the complex structures of supersymmetry. They allow physicists to visualize the transformations and interactions of supersymmetric particles, aiding in the development of new theories and models.

- **Bridging Mathematics and Physics:** Gates' work exemplifies the deep connection between advanced mathematics and fundamental physics. By using graphical symbols like Adinkras, he bridges abstract mathematical concepts with tangible physical theories, enhancing our ability to explore and conceptualize high-dimensional spaces and symmetries.

- **Enhancing Computational Models:** The structural clarity that Adinkras provides has implications for computational physics. They enable more efficient modeling and simulation of supersymmetric theories, which is crucial for testing these theories against experimental data and for exploring their cosmological and particle physics implications.

- **Educational Impact:** Beyond their research value, Adinkras serve as educational tools, helping students and newcomers to the field grasp the foundational concepts of supersymmetry and supergravity more intuitively. This has the potential to inspire and train the next generation of physicists.

Challenges and Future Directions

Professor Gates' work on Adinkras and supersymmetry has been groundbreaking new directions for future research:

- **Empirical Verification:** The Large Hadron Collider (LHC) and future particle accelerators continue to search for supersymmetric particles. Gates and others provide the theoretical foundation needed to understand what these particles might look like and how they might be detected.

- **Further Mathematical Development:** The field of Adinkra research is still evolving. There is much to be explored in terms of the mathematical properties of these symbols and how they can be generalized or extended to encompass even more complex aspects of supersymmetry and supergravity.

- **Interdisciplinary Applications:** The potential applications of Adinkra symbols extend beyond physics. In computer science, for example, similar graphical representations are used in the study of algorithms and data structures. Exploring these interdisciplinary connections could lead to novel computational techniques inspired by supersymmetric principles.

Conclusion

Professor Sylvester James Gates, Jr.'s pioneering work on Adinkra codes and his contributions to the fields of supersymmetry and supergravity have significantly advanced our understanding of fundamental physics. His innovative approach to visualizing complex mathematical structures has deepened our grasp of theoretical physics and opened new pathways for interdisciplinary research and education. As we continue to explore the universe at its most fundamental levels, the work of Gates and his contemporaries will undoubtedly play a crucial role in shaping our quest for a unified understanding of the cosmos.

Professor Sylvester James Gates, Jr.'s and his team's observation that error-correcting codes, similar to those used in search engines and web browsers, are embedded in the fabric of the universe is a profound insight that bridges the seemingly disparate worlds of fundamental physics and information technology. This statement hints at a deep, underlying structure in the universe that mirrors the constructs devised by humans to ensure the reliability and efficiency of digital communications.

The Nature of Error-Correcting Codes

Error-correcting codes are algorithms or sequences used in computing and telecommunications to detect and correct errors within data transmissions. In digital systems, these codes are crucial because they allow systems to identify and fix errors that arise due to noise, signal degradation, or interference, thereby ensuring that the information transmitted from one point to another is accurate and intact.

In technology, these codes operate under algorithms that add redundancy to the original data. For instance, in a simple parity-check system, an extra bit is added to a string of binary data to indicate whether the sum of all the 1s in the string is odd or even. More sophisticated systems, like Reed-

Solomon or Turbo Codes, employ complex mathematical structures to detect and correct multiple errors over large data blocks.

The answer to how these codes could appear in the fundamental laws of physics given by Prof. Gates diverges from the simulation hypothesis (that requires programmers, either natural or supernatural) but is no less radical.

The only place in the natural world where evidence for error-correcting codes has been noted is in the area of genetics. In this domain, it has been argued that the 'engine' of natural selection is the source. Using this as a model, his proposal that the classical error-correcting codes (CECCs) discovered in the fundamental mathematical laws for supersymmetry (SUSY) must have occurred by an evolutionary law active on the laws of SUSY mathematics.

Implications for Fundamental Physics

When Professor Gates talks about finding similar error-correcting codes within the fundamental equations of the universe, particularly in the context of string theory and supersymmetry, it suggests that the universe at its most elemental level may operate under principles akin to those used in digital communications and data processing.

- **Universality of Information:** This discovery points to the idea that information and its integrity might be fundamental aspects of physical reality. Just as error-correcting codes preserve the accuracy of data in a digital system, similar mechanisms at the quantum or cosmic level could be responsible for maintaining the stability and consistency of the laws of physics across the vast expanse of the universe.

- **Quantum Gravity and Error Correction:** In the realm of quantum gravity, researchers have hypothesized that space-time itself may be quantized. If true, the presence of error-correcting codes suggests that the quantum fabric of space-time might have built-in mechanisms to prevent informational degradation or loss. This could be a crucial feature that allows for the coherent evolution of the universe and might explain why certain physical laws appear so finely tuned and stable over cosmological timescales.

- **Digital Physics and the Simulation Hypothesis:** The discovery also fuels speculative theories such as the digital physics hypothesis, which suggests that the universe could be akin to a giant computational system, or even the simulation hypothesis, which posits that reality might be a sophisticated virtual simulation. While these ideas are more philosophical than scientific, the presence of error-correcting codes adds a layer of intrigue, suggesting that the universe's deepest laws might mirror the algorithms designed by humans for digital and computational purposes.

- **Theoretical Implications for Unified Theories:** For string theory and theories of everything (ToEs), the presence of error-correcting codes could indicate a pathway to deeper unification. These codes might provide a clue to how different forces and particles are integrated at the quantum level, potentially offering insights into how to merge quantum mechanics with general relativity or how to predict the behavior of yet-undiscovered fundamental particles.

Future Directions and Challenges

This fascinating intersection of digital principles and fundamental physics opens several avenues for research and exploration:

- **Empirical Validation:** One of the biggest challenges is finding empirical evidence for the presence and role of error-correcting codes in fundamental physics. This would require innovative experiments and observations that could detect the subtle ways these codes manifest in the behavior of particles and fields.

- **Theoretical Development:** Theoretically, physicists need to explore how these codes are integrated into the mathematics of string theory, quantum field theory, and other frameworks. This involves a blend of information theory, quantum mechanics, and mathematics to formulate a coherent model that explains the presence and functionality of these codes.

- **Information Theory and Physics:** The relationship between information theory and physics is a growing field of study. Researchers need to delve deeper into how concepts from information theory, such as entropy, information, and error correction, can be applied to understand the physical universe more comprehensively.

Conclusion

In summary, Professor Sylvester James Gates, Jr.'s discovery of error-correcting codes in the fabric of the universe is a groundbreaking insight that blurs the lines between information technology and fundamental physics. It suggests a universe that is not only governed by physical laws but also by principles akin to those that govern data and information. This perspective could lead to a deeper understanding of the cosmos, revealing it not just as a physical reality but as an informational structure where the preservation and processing of data play a central role in its evolution and stability.

Adinkra Symbols of the Dogon Tribe

The Dogon tribe, an ethnic group indigenous to Mali in West Africa, is renowned for its rich cultural heritage and profound astronomical knowledge, particularly concerning the star system of Sirius. The Dogon have an intricate cosmology that includes detailed information about Sirius A, its companion stars Sirius B and the hypothetical Sirius C, and the planets within our solar system. What is astonishing about the Dogon's astronomical knowledge is that they possessed intricate details about these celestial bodies long before modern science confirmed their existence.

Sirius A is the brightest star in the night sky, and for the Dogon, it holds a central place in their cosmology. However, the truly remarkable aspect of their knowledge pertains to Sirius B. This white dwarf star is invisible to the naked eye and was only discovered by Western astronomers in 1862.

The Dogon not only acknowledged the existence of Sirius B but also described its incredibly dense nature and its elliptical orbit around Sirius A with a precision that aligns closely with modern astronomical findings. Additionally, there are references in Dogon mythology to a third star, Sirius C, which remains a subject of debate among astronomers.

According to Dogon mythology, this advanced astronomical knowledge was imparted to them by the Nommo, a race of amphibious beings who descended from the stars. The Nommo are described as ancestral spirits or extraterrestrial beings who came to Earth in an ark, bringing with them wisdom and knowledge. The Dogon recount that the Nommo shared secrets of the universe, including the existence of Sirius B and the complex structure of the cosmos. This narrative suggests a profound connection between the Dogon and celestial phenomena, positioning the Nommo as both cultural heroes and bearers of esoteric knowledge.

The Dogon's understanding of the planets in our solar system is also intriguing. They possess detailed descriptions of Jupiter's four major moons and Saturn's rings, information that aligns closely with observations made through modern telescopes. This level of astronomical sophistication, especially when considering the tools available to the Dogon at the time, has led to considerable speculation and fascination. Some researchers propose that the Dogons' knowledge could have been acquired through ancient, now-lost advanced civilizations or through direct contact with beings, which they called the Nommo, which came from Sirius B, as their star lore suggests. They knew of the existence of a white dwarf named Sirius B. American astronomer and telescope maker Alvan Graham Clark discovered Sirius B on January 31, 1862. Clark was testing a new telescope on Sirius when he noticed the faint companion star, 10,000 times dimmer than Sirius and nicknamed "the Pup." Sirius B is so faint that it's almost lost in Sirius's glare, which is why it escaped detection until then. This star cannot be seen with the naked eye.

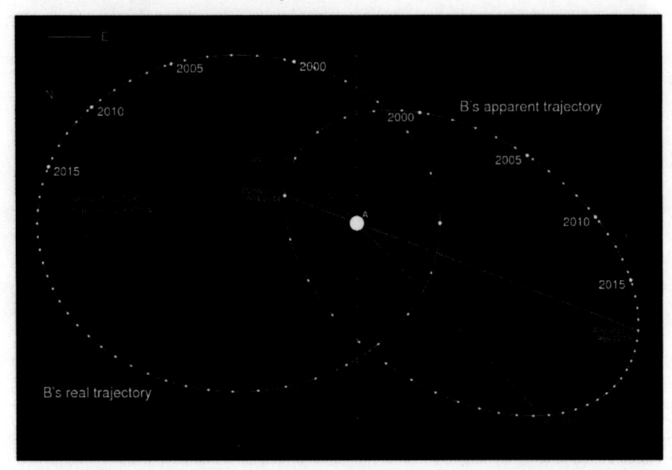

The orbit of Sirius B around A, as seen from Earth (slanted ellipse). The wide horizontal ellipse shows the true shape of the orbit (with an arbitrary orientation) as it would appear if viewed straight on. - Wikipedia

The Dogon tribe's astronomical knowledge stands as a compelling testament to the rich and complex understanding of the universe held by ancient cultures. Whether through oral traditions, lost ancient technology, or extraordinary encounters with otherworldly beings, the Dogon's precise knowledge of Sirius and our solar system continues to inspire wonder and scholarly investigation.

The Dogon Adinkra symbols and Professor James Gates, Jr.'s Adinkra codes share intriguing similarities, particularly when considered as three-dimensional objects, sparking interest and speculation in both scientific and cultural contexts.

Dogon Adinkra Symbols

The Dogon people, an ethnic group indigenous to Mali in West Africa, are known for their rich cultural heritage and symbolic representations. Their Adinkra symbols are visual motifs used to convey proverbs, concepts, and traditional wisdom. These symbols are often found in textiles, pottery, and other forms of art.

Professor James Gates, Jr.'s Adinkra Codes

Professor Sylvester James Gates, Jr., an American theoretical physicist, introduced a concept he named "Adinkra codes" in the context of supersymmetry and string theory. These codes are mathematical representations that use binary matrices and graphs to describe the algebraic structure of supersymmetric theories. The term "Adinkra" was chosen due to the visual resemblance of these codes to traditional African Adinkra symbols, though Gates' usage is purely mathematical.

Similarities in Three-Dimensional Representations

- **Geometric Complexity:** Both sets of symbols exhibit complex geometric patterns. The traditional Dogon symbols, when visualized in three dimensions, form intricate shapes that resemble polyhedral structures. Similarly, Gates' Adinkra codes, when expanded into three dimensions, form complex, symmetrical shapes reminiscent of polyhedra and higher-dimensional objects.

- **Symmetry and Aesthetics:** The symmetry observed in both sets of symbols is striking. Traditional Dogon symbols often incorporate symmetrical patterns, which are aesthetically pleasing and carry symbolic meaning. Gates' Adinkra codes also exhibit high levels of symmetry, which is crucial for the mathematical properties they represent in supersymmetry.

- **Cultural and Mathematical Resonance:** The visual and structural resonance between these symbols raises fascinating questions about the universality of certain patterns and shapes across different fields and cultures. It suggests a potential underlying connection between the ways humans conceptualize and represent complex ideas, whether through cultural symbolism or advanced mathematical theory.

Speculative Connections

While the similarities are compelling, it's important to note that there is no established historical or cultural link between the Dogon symbols and Gates' mathematical codes. The connection could be a fascinating coincidence, highlighting how certain patterns and structures can recur in different contexts.

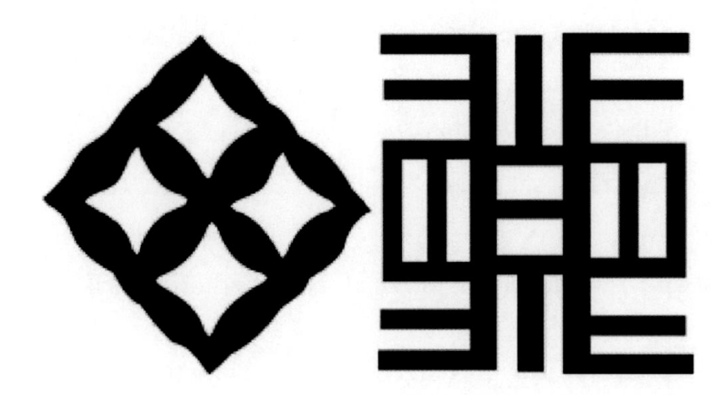

Adinkra Symbols – Dogon Tribe – Mali Africa - umich.edu

- **Cultural Inspiration:** Gates has mentioned that his choice of the term "Adinkra" was influenced by the visual similarity to the traditional African symbols, though the mathematical properties of his codes are not derived from these cultural symbols.

- **Universal Patterns:** The recurrence of similar patterns in both cultural and scientific contexts might suggest that certain geometric shapes and symmetries are universally appealing or significant. This could reflect underlying principles of human cognition, perception, and the natural world.

In summary, the eerie similarity between Dogon Adinkra symbols and Professor James Gates, Jr.'s Adinkra codes, when expanded to three-dimensional objects, serves as a fascinating example of how complex patterns can transcend different domains, from cultural heritage to advanced theoretical physics.

How Holograms Work and Their Connection to Fractals

Holography, the science and art of creating holograms, is a fascinating field that combines physics, mathematics, and visual science to produce three-dimensional images that appear strikingly real and tangible. This complex interaction of light, materials, and observation creates not just stunning visuals but also embodies profound principles, including those related to fractals, the self-similar patterns found throughout nature and mathematics.

Understanding Holography: The Science of Three-Dimensional Imaging

A hologram is essentially a photographic recording of a light field used to display a fully three-dimensional image of the holographic subject. Unlike traditional photography, which captures a two-dimensional image, holography captures the light scattered from an object and reconstructs it so that it appears as if the object is still there, even when it is not.

- **The Role of Light and Interference:** At its core, holography is about capturing the interference pattern between two sets of light waves—a reference beam and an object beam. The reference beam is a part of the coherent light (usually from a laser) that directly illuminates the recording medium, such as photographic film or a digital sensor. The object beam is the light that reflects off the object being holographic and then strikes the recording medium.

- **Creating the Hologram:** When these two beams meet at the recording medium, they create an interference pattern due to the way the peaks and troughs of the light waves combine. This pattern encodes not just the intensity of the light, as a photograph would, but also its phase – the position of the peaks and troughs. This phase information allows the hologram to reconstruct the light waves as if they were emanating from the original object when viewed later.

- **Viewing the Hologram:** To view a hologram, the recorded interference pattern is illuminated with a reconstruction beam, usually from the same type of laser used to create it. This light interacts with the hologram to produce a light field identical to the one originally reflected from the object. As a result, the viewer sees a three-dimensional image that changes perspective and depth in response to their movements, just as if the original object were still present.

Holograms and Fractals: Exploring the Connection

While holography is a well-established field in optics, its connection to fractals is a more recent exploration, illuminating how these self-similar patterns can influence and enhance holographic imaging.

- **Fractal Nature of Light Waves:** At a fundamental level, light waves themselves can exhibit fractal-like behavior under certain conditions. The wave patterns created by interference have been shown to have fractal characteristics, particularly in complex holographic setups where multiple reflections and diffractions occur. This suggests that the light field itself can be thought of as a fractal structure in certain contexts.

- **Fractal Holograms:** Researchers have developed methods to create holograms based on fractal designs. These fractal holograms use the self-similar patterns of fractals to modulate the phase and amplitude of light in a way that traditional geometric patterns cannot. This approach can produce holograms with unique optical properties, such as the ability to project images at multiple scales or depths simultaneously, leveraging the inherent scaling properties of fractals.

- **Enhanced Image Resolution and Depth:** The fractal approach to holography can enhance the resolution and depth of the holographic image. Because fractals contain structures at many scales, they can help encode more information into the hologram's interference pattern. This leads to a more detailed reconstruction with greater depth of field, as the fractal patterns can influence the way light reconstructs the image at different viewing angles and distances.

- **Applications in Data Storage and Security:** Beyond imaging, the fractal properties of holograms have implications for data storage and security. Fractal holograms can store data at multiple scales, increasing the density of information storage. Moreover, the complex patterns of fractal holograms make them difficult to replicate or forge, offering a new avenue for secure holographic watermarking and anti-counterfeiting measures.

Challenges and Future Directions

While the fusion of holography and fractals opens new horizons, it also presents challenges and opportunities for further research.

- **Complexity in Creation and Computation:** Designing and creating fractal holograms requires sophisticated computational models and precise control over the light-modulating elements. Advances in computational holography and the development of new materials for dynamic hologram creation are critical to fully harnessing the potential of fractal holography.

- **Interdisciplinary Research:** Exploring the connection between fractals and holography requires an interdisciplinary approach, combining insights from physics, mathematics, computer science, and art. This collaboration can lead to innovative designs and applications, from enhanced virtual reality experiences to new forms of artistic expression.

- **Educational and Practical Implications:** The study of fractal holography not only advances our understanding of light and perception but also has practical implications. For instance, it can improve medical imaging techniques, enhance optical communications, and revolutionize display technologies.

Conclusion

The intersection of holography and fractals represents a vibrant field of study that melds intricate mathematics with practical applications in imaging and beyond. By understanding and harnessing the fractal properties of light and materials, scientists and engineers can create more detailed, secure, and information-rich holograms, opening up new possibilities for exploring the nature of light and perception. As this field evolves, it promises to reveal further wonders of the fractal universe hidden within the beams of light that illuminate our world.

As we delve into the depths of matter at its smallest scale, an intriguing pattern emerges: the whole is reflected in its smallest parts, and what diminishes is not the essence but merely the resolution. This observation suggests that we might be inhabiting a fractal holographic matrix, where each fragment, no matter how minuscule, contains a blueprint of the entire structure. In such a universe, the fractal nature ensures that every level of complexity is a microcosm of the larger structure, while the holographic principle posits that the entirety of the cosmos can be inferred from its smallest segments. This perspective not only challenges our conventional understanding of space and time but also hints at a profound interconnectedness, suggesting that the very fabric of reality is woven with patterns that repeat across all scales, seamlessly connecting the macroscopic and microscopic worlds.

Harnessing the Power of Fractals in Technology: From Movie Animation to AI Video and Image Creation

Fractals, with their intricate self-similar patterns and ability to describe complex natural phenomena, have transcended their mathematical origins to become a powerful tool in various technological applications. From the stunning visuals in Pixar movies to the cutting-edge developments in artificial intelligence (AI) for video and image creation, fractals are revolutionizing the way we perceive and create digital content.

Fractals in Movie Animation: The Pixar Revolution

One of the most visible applications of fractals in technology is in the field of movie animation, particularly in the work of studios like Pixar. These studios have harnessed the power of fractals to create realistic and enchanting environments that captivate audiences worldwide.

- **Creating Natural Landscapes:** In animation, the challenge is to create scenes that are both visually appealing and authentic to the natural world. Fractals are perfect for this task because they mimic the way nature organizes itself. For instance, the fractal geometry of mountains, trees, and clouds can be modeled to produce landscapes that have the depth and irregularity seen in real life. Pixar's "Brave" is an excellent example of how fractal algorithms were used to render the complex, lush landscapes of ancient Scotland, providing a rich backdrop that feels both magical and authentic.

- **Detailing and Complexity:** Fractals allow animators to introduce an incredible level of detail into their scenes without manually modeling every aspect. By using fractal-based algorithms, studios can generate complex patterns like the roughness of tree bark, the distribution of leaves, or the flow of water, all of which add to the realism and richness of the animation. In "Finding Nemo," fractals were used to simulate the movement of water and the way light interacts with it, creating a believable underwater world that is both complex and captivating.

- **Efficiency and Scalability:** Using fractals in animation is not just about aesthetic appeal; it's also about efficiency. Fractals allow for the creation of highly detailed scenes using relatively simple mathematical rules, which can be scaled up or down without losing fidelity. This means that animators can create vast, detailed worlds with fewer resources and less manual labor, optimizing both time and cost in film production.

AI Video and Image Creation: The Frontier of Fractal Technology

In the realm of AI, fractals are finding new applications, particularly in video and image creation, where they contribute to advancements in both the generation and enhancement of digital media.

- **Generative Adversarial Networks (GANs):** AI models, particularly GANs, have been used to create complex images and videos from scratch. Fractal principles are integrated into these models to generate textures and patterns that mimic those found in nature. For instance, AI can use fractal algorithms to create realistic skin textures, cloud formations, or even entire landscapes for virtual environments. This not only enhances the visual quality of AI-generated images and videos but also ensures that they maintain a level of randomness and variety that prevents repetitive or artificial looks.

- **Image Enhancement and Upscaling:** Fractals are also instrumental in image enhancement and upscaling processes. Techniques like fractal compression take advantage of the self-similarity in images to reduce file size without significant loss of quality. Similarly, when upscaling images or videos, fractal algorithms can predict and fill in missing details by analyzing the patterns in the available data, thereby improving resolution while maintaining the integrity of the original imagery.

- **Deep Learning and Fractal Architecture:** The structure of neural networks themselves can be influenced by fractal concepts. Fractal neural networks, which mimic the fractal organization of natural systems, can create more efficient and robust AI models.

These networks use fractal designs to optimize information flow and processing, leading to improvements in how AI understands and generates complex visual content.

Real-World Applications and Future Directions

The use of fractals in technology extends beyond animation and AI, influencing various fields and suggesting exciting future developments.

- **Virtual Reality (VR) and Augmented Reality (AR):** In VR and AR, fractals can be used to create detailed and immersive environments that require minimal computational resources. As these technologies continue to evolve, fractal algorithms will play a crucial role in rendering realistic virtual worlds where users can interact with complex, dynamic elements in real time.

- **Medical Imaging:** Fractals are used in medical imaging to analyze patterns in tissues and organs. This analysis can help identify abnormalities like tumors or fractures, which often disrupt the normal fractal patterns of healthy tissues. By harnessing the power of fractals, medical professionals can improve diagnostic accuracy and tailor treatments more effectively.

- **Environmental Modeling:** In environmental science, fractals are used to model complex phenomena like weather patterns, the spread of wildfires, or the growth of ecosystems. These models help scientists predict changes and plan interventions more accurately, ensuring better management of natural resources and disaster response.

- **Architecture and Design:** Architects and designers use fractals to create structures and products that are not only aesthetically pleasing but also structurally sound and efficient. The use of fractal geometry in designing buildings, furniture, or even textiles can lead to innovations that combine beauty with functionality.

Conclusion

In conclusion, the power of fractals is being harnessed across various domains of technology, from movie animation and AI to VR and medical imaging. This mathematical concept, rooted in the patterns of nature, has found its way into the digital world, where it enhances realism, efficiency, and creativity. As technology continues to evolve, the role of fractals is set to expand, promising new ways to create, analyze, and interact with the digital and physical worlds.

The concept of creating immersive, realistic environments, as epitomized by the advanced holodeck in "Star Trek" or modern-day flight simulators, shares striking similarities with the theoretical idea of a fractal holographic universe. Both notions revolve around constructing detailed, interactive worlds that can serve as powerful tools for humanity, facilitating learning, exploration, and growth in ways that blend the boundaries between reality and simulation.

Simulated Realities: Holodecks and Flight Simulators

The holodeck from "Star Trek" is a fictional advanced simulation technology that creates fully immersive, interactive environments indistinguishable from reality. This technology allows users to engage with simulated scenarios for entertainment, training, or research purposes. Similarly, modern-day flight simulators are sophisticated systems used to train pilots, providing realistic flight experiences without the risks and costs associated with actual flying. These simulators use advanced graphics, motion platforms, and other sensory inputs to replicate the experience of flying as accurately as possible.

Both these technologies rely on detailed modeling of the real world to create their immersive experiences. They use algorithms and data to generate environments, behaviors, and responses that mimic those found in natural settings. The accuracy and effectiveness of these simulations are critical, as they directly impact the user's learning and decision-making processes.

Fractal Holographic Universe: A Grand Simulation

Expanding this concept to the idea of a fractal holographic universe suggests a parallel where our entire universe could be akin to a vast, intricate simulation. In this scenario, the universe is viewed as a holographic projection, with fractal patterns ensuring that every part contains information about the whole. This concept aligns with theories in physics and cosmology that suggest the universe's fundamental structure might be holographic and fractal in nature.

- **Educational and Experiential Learning:** Just as holodecks and flight simulators are used for training and education, a fractal holographic universe could be seen as a grand educational tool for humanity. In this cosmic "simulation," individuals learn through direct experience, exploring the laws of physics, the complexities of ecosystems, and the nuances of human behavior. Each experience in this universe would be a lesson in the interconnectedness and dynamics of reality, much like how pilots learn to handle various flight conditions in a simulator.

- **Exploration and Problem Solving:** In "Star Trek," the holodeck is often used for problem-solving, allowing crew members to test different scenarios and strategies in a controlled environment. Similarly, if our universe is a fractal holographic construct, it could be perceived as a platform for testing ideas, exploring new concepts, and solving problems on a cosmic scale. The fractal nature ensures that solutions and insights gained at one level can be applicable across various scales, enhancing our ability to innovate and adapt.

Disney's HoloTile Floor is an omnidirectional treadmill floor that allows users to move in any direction while in a VR environment. The floor is made up of hundreds of small, hexagonal plates that can tilt and angle individually to simulate natural movement. The floor uses Lidar-based sensors, custom software, and rotating disks to move subtly underfoot and create the illusion of walking.

The HoloTile Floor can move any person or object, like telekinesis, and multiple people can be on it, walking independently. It can be used for immersive VR and AR experiences, and one clear use is virtual reality, as users aren't in danger of walking into a couch or rail while wearing a headset. Users don't need to wear any special equipment to run, jump, and move naturally in any direction.

HoloTile Floor by Disney – DigitalTrends.com

Disney's HoloTile Floor represents a cutting-edge innovation in interactive flooring technology, leveraging holographic projections to create dynamic and engaging environments. This technology opens up numerous possibilities for various applications beyond just entertainment. Here's an exploration of the types of holodeck and other technologies that could utilize HoloTile floors, along with potential applications.

Types of Technologies Utilizing Holo Tile Floors

1. Holodecks and VR Spaces

Interactive Holodecks:

- **Description:** Holodecks, popularized by science fiction series like Star Trek, are rooms where environments and experiences are simulated using holography and other immersive technologies.
- **Application:** HoloTile floors can enhance holodeck experiences by providing dynamic, interactive surfaces that change in response to user movements. This allows for more realistic simulations where the floor can mimic different terrains and environments in real time.

Here is the image of a person standing inside a holodeck similar to the one in Star Trek

Immersive VR Spaces:

- **Description:** VR spaces designed for immersive experiences, training, and entertainment.

- **Application:** In combination with VR headsets, HoloTile floors can provide tactile feedback and interactive visuals that enhance the sense of presence and immersion, making users feel as if they are truly walking on different surfaces.

2. Augmented Reality (AR) Environments

AR Exhibits

- **Description:** AR environments that overlay digital content onto the real world.

- **Application:** HoloTile floors can serve as interactive canvases that display AR content, such as historical reconstructions, educational information, or gaming elements, which respond to user interactions and movements.

Retail and Showrooms

- **Description:** AR-enhanced retail spaces and showrooms that provide interactive product displays and experiences.

- **Application:** Retailers can use HoloTile floors to create engaging shopping experiences where customers can see product information, promotions, and virtual try-ons projected onto the floor as they move through the store.

3. Interactive Gaming and Entertainment

Interactive Gaming Floors

- **Description:** Gaming environments that use interactive surfaces to engage players in physical and virtual gameplay.

- **Application:** HoloTile floors can be used to create interactive games where the floor itself becomes part of the gameplay, displaying dynamic terrains, obstacles, and game elements that respond to player actions.

Themed Attractions

- **Description:** Attractions in theme parks or entertainment centers that offer immersive experiences.

- **Application:** HoloTile floors can transform theme park attractions by creating interactive pathways that change to match different themes or stories, enhancing the overall visitor experience.

4. Educational and Training Environments

Interactive Classrooms

- **Description:** Classrooms equipped with technology to provide interactive and engaging learning experiences.

- **Application:** HoloTile floors can be used in educational settings to create interactive lessons where the floor displays maps, diagrams, and interactive quizzes, making learning more engaging and hands-on.

Simulation-Based Training

- **Description:** Training environments that use simulations to teach complex skills.
- **Application:** Industries like aviation, medicine, and the military can use HoloTile floors in their training simulations to create realistic scenarios that require trainees to navigate through dynamic, changing environments.

5. Healthcare and Rehabilitation

Therapeutic Environments

- **Description:** Spaces designed to aid in therapy and rehabilitation.
- **Application:** HoloTile floors can be used in physical therapy and rehabilitation to create interactive exercises and scenarios that help patients recover mobility and balance in a stimulating and motivating environment.

Virtual Rehabilitation

- **Description:** Rehabilitation programs that use virtual environments to assist in recovery.

- **Application:** Patients can walk on HoloTile floors that simulate different terrains and environments, providing varied and challenging exercises tailored to their rehabilitation needs.

Potential Applications of HoloTile Floors

Museums and Exhibitions

- **Enhanced Exhibits:** Museums can use HoloTile floors to create interactive exhibits where the floor displays historical footage, artifacts, or geographical maps that visitors can interact with.

Corporate Environments

- **Dynamic Workspaces:** Corporate offices can integrate HoloTile floors in conference rooms and common areas to display interactive presentations, company information, or dynamic decor that changes based on meetings and events.

Public Spaces

- **Engaging Installations:** Public spaces like plazas and airports can use HoloTile floors to provide information, entertainment, and navigation assistance to visitors, making these spaces more engaging and informative.

Art Installations

- **Interactive Art:** Artists can use HoloTile floors to create dynamic art installations that change and respond to viewer movements, creating a more immersive and interactive art experience.

Conclusion

Disney's HoloTile Floor technology has the potential to revolutionize a wide range of industries by providing dynamic, interactive surfaces that enhance user experiences. From immersive holodecks and VR spaces to educational environments and healthcare applications, the possibilities are vast. This technology not only enriches entertainment and learning but also offers practical benefits in training, rehabilitation, and public engagement, demonstrating the transformative power of interactive, holographic floors.

Personal and Collective Growth: Flight simulators provide pilots with a safe space to hone their skills, learn from mistakes, and improve their performance. Extending this to the fractal holographic universe, one could argue that life experiences—both individual and collective—are opportunities for growth and evolution. The challenges and trials faced by humanity can be viewed as "simulations" within this greater framework, designed to foster development, resilience, and a deeper understanding of our place in the cosmos.

Interactivity and Adaptability: A key feature of both the holodeck and flight simulators is their interactivity and ability to adapt to the user's actions and decisions. This dynamic interplay is essential for effective learning and exploration. In the context of a fractal holographic universe, this suggests that the universe is responsive to human actions and thoughts, adapting and evolving in ways that reflect our collective beliefs, decisions, and behaviors.

Implications and Future Directions

The analogy between advanced simulation technologies and a fractal holographic universe opens up profound philosophical and scientific questions about the nature of reality and our role within it.

- **Technological Advancements:** As we continue to develop more sophisticated simulation technologies, we might get closer to creating environments that are indistinguishable from the "real" world. This progression could provide insights into how a fractal holographic universe might operate and interact with its inhabitants.

- **Scientific Exploration:** The idea encourages scientists to explore the holographic and fractal properties of the universe more deeply, potentially leading to breakthroughs in understanding quantum gravity, dark matter, and the early universe.

- **Philosophical Reflections:** This perspective invites philosophical reflections on the nature of existence, consciousness, and the potential for humanity to transcend its current limitations by harnessing the power of simulations for growth, learning, and exploration.

Conclusion

In conclusion, the parallels between advanced simulation technologies like holodecks and flight simulators and the concept of a fractal holographic universe highlight the potential of simulations as tools for understanding, learning, and evolving. By exploring these connections, we not only gain insights into the nature of our universe but also open new avenues for technological innovation and philosophical inquiry.

Chapter 10

Spirituality, Philosophy, and the Fractal Universe

Ancient Wisdom and the Recognition of Unity: Understanding the Nature of Our Reality

THROUGHOUT HISTORY, VARIOUS CULTURES have explored profound questions about the nature of reality, consciousness, and the cosmos. Many ancient philosophies and spiritual traditions propose that the world we experience is not the ultimate reality but a transient, illusory stage. This perspective, often described as the concept of life being a "dream," suggests that upon death, we "wake up" to a truer, more unified state of being. These ancient insights provide a fascinating lens through which to view the interconnectedness and unity that underpin our existence.

The Nature of Reality in Ancient Wisdom

- **Hinduism and the Concept of Maya:** In Hindu philosophy, particularly in Vedanta, the world is often described using the concept of "Maya," which translates to "illusion" or "magic." According to this view, the physical universe and all its experiences are a kind of cosmic illusion, where the ultimate reality is Brahman, the singular, unchanging truth behind all forms and names. The Upanishads, ancient texts of Hindu philosophy, teach that individual souls (Atman) are essentially one with Brahman, but they are caught in the cycle of birth and rebirth (samsara) due to ignorance of their true nature. Enlightenment, or moksha, is the realization of this oneness and the liberation from the illusion.

- **Buddhism and the Doctrine of Anatta:** Buddhism takes a somewhat different approach but arrives at a similar conclusion about the illusory nature of the self and the world. Through the doctrine of "Anatta" (no-self), Buddhism asserts that what we consider the self is merely a collection of temporary aggregates (skandhas) and that clinging to this false sense of self is the source of suffering. The Buddhist path aims to awaken individuals from the dream-like state of ignorance to the realization of Sunyata (emptiness) and interconnectedness of all things, where distinctions between self and other dissolve.

- **Taoism and the Unnamable Tao:** Taoist philosophy, as expressed in the "Tao Te Ching" by Laozi, speaks of the Tao, the fundamental, ineffable principle that underlies and unites all aspects of the universe. The Tao is beyond words and concepts, and the sage who understands this lives in harmony with its spontaneous flow. In this worldview, the apparent separateness of things is an illusion, and true wisdom lies in recognizing the unity and interdependence of all life.

- **Indigenous and Shamanic Traditions:** Many indigenous and shamanic cultures, such as the Aboriginals, hold a view of reality where the physical world is a shadow or reflection of a more profound spiritual reality. In these traditions, life is often seen as a dream, and death is a transition to another state of consciousness where the deeper truths of existence are revealed. These cultures use rituals, visions, and meditative practices to connect with this larger reality and gain insights into the nature of existence and the unity of all beings.

Aboriginal Elder – 4biddenknowledge Inc

Interconnectedness and Unity in Ancient Wisdom

The idea of unity—that all things are intrinsically connected and that individual identity is a temporary, illusory construct—is central to many ancient wisdom traditions. This understanding has profound implications for how we perceive ourselves and our place in the cosmos.

- **The Illusion of Separateness:** Ancient wisdom teaches that the sense of separateness—between self and other, between humanity and nature—is a fundamental misunderstanding. By recognizing that all is one, individuals can overcome egoic barriers, reduce suffering, and foster a deeper sense of compassion and empathy for all life forms.

- **The Role of Consciousness:** In many traditions, consciousness is not confined to individual beings but is seen as a universal attribute that permeates all of existence. This shared consciousness is the fabric that binds the universe together, and becoming aware of this unity is key to spiritual awakening and enlightenment.

- **The Cycle of Birth and Rebirth:** The concept of reincarnation, found in Hinduism, Buddhism, and other belief systems, posits that the soul undergoes cycles of birth and rebirth, each time learning and evolving until it realizes its true nature and breaks free from the cycle. This journey is a process of gradually awakening from the dream of material existence to the reality of unity and interconnectedness.

- **Mystical Experiences and the Perception of Reality:** Mystical experiences, whether through meditation, prayer, or entheogenic substances, often lead to profound realizations about the unity of all things. These experiences can dissolve the boundaries of the self and reveal a reality where everything is interconnected in a vast, living web.

Contemporary Implications and Reflections

The ancient perspectives on the dream-like nature of reality and the promise of awakening have significant implications for contemporary life and thought.

- **Environmental Consciousness:** Understanding the interconnectedness of all life can lead to greater environmental awareness and responsibility. If we see the Earth and all its inhabitants as part of a unified whole, the motivation to protect and preserve the planet becomes a natural extension of self-care.

- **Social and Ethical Behavior:** The recognition of unity challenges us to reconsider our ethical frameworks. If all beings are interconnected, then harm done to one is harm done to all. This realization can foster a more compassionate, inclusive approach to social justice and human rights.

- **Personal Growth and Mental Health:** The idea that life is a dream from which we can awaken can offer new perspectives on personal challenges and mental health issues. By understanding the impermanent, illusory nature of many of our struggles, individuals can find pathways to healing and transformation that transcend conventional approaches.

- **Scientific and Philosophical Inquiry:** The ancient wisdom on unity and the dream-like nature of reality invites scientists and philosophers to explore the intersections between consciousness, physical reality, and the nature of the universe. This exploration could lead to new paradigms in fields like quantum physics, cosmology, and the study of consciousness.

Conclusion

In conclusion, the ancient wisdom about the dream-like nature of reality and the unity of all existence offers profound insights into the nature of our being and the cosmos. By reflecting on these teachings and incorporating their essence into our lives and thoughts, we can foster a deeper understanding of ourselves, cultivate compassion and empathy for others, and move closer to realizing the interconnected, unified nature of all that is.

Exploring Profound Insights from The Matrix Movie:

The movie "The Matrix" offers a wealth of thought-provoking dialogue, particularly when we delve into the concepts presented by #Morpheus. Let's examine some key excerpts.

Morpheus on the Nature of Reality

"What is real? How do you define 'real'?" Morpheus asks. If "real" is what we can feel, smell, taste, and see, then it's essentially just electrical signals interpreted by our brains. This perspective challenges our basic understanding of reality, suggesting that our sensory experiences might not be as straightforward as they seem.

The Illusion of Physicality

Morpheus questions the physicality of the Matrix world: "Do you believe my being stronger or faster has anything to do with my muscles in this place? You think that's air you're breathing?" These rhetorical questions prompt reconsidering what we perceive as physical limitations or necessities within a constructed reality.

Neo's Realization and the Illusion of the World

As Neo begins to grasp that the interpretation of these signals defines his perception of being awake or asleep, he understands that everything within the dream world is an illusion. This realization paves the way for a new paradigm shift. Like each of us, Neo is influenced by a thought structure built upon past experiences. He must unlearn what he has learned, becoming mindful of how his past experiences color his present and learning to detach from these preconceptions to truly free his mind.

The Power of Belief and the Chains of Perception

The movie further explores how our beliefs bind the world. It posits that the key to salvation lies within oneself. #Belief is a potent force; the thoughts we harbor are powerful, and illusions can have as profound an impact as the truth. A madman, for instance, perceives his delusions as reality, unshakeable in his conviction. It is only when the origin of these thoughts is questioned that the possibility of liberation emerges.

Transformation Through Changing One's Mind

The path to salvation is presented as a simple yet profound shift – changing one's mind. Altering one's mindset leads to a transformation in the source of all thoughts, ideas, and perceptions, both past and future. This shift liberates one from previous misconceptions and opens up new possibilities for future discovery. By changing the source of your thoughts, you free yourself from old patterns and open up a new path for understanding and perception.

In summary, "The Matrix" presents a deep exploration of reality, perception, and the power of the human mind. It challenges viewers to reconsider their understanding of the world, the influence of their beliefs, and the potential for change that lies within each individual.

Professor Sylvester James Gates, Jr., a renowned theoretical physicist, currently serving as the John S. Toll Professor of Physics at the University of Maryland and as the Director of The Center for String Theory & Particle Theory, has made a groundbreaking discovery in the field of string theory. His work on supersymmetrical equations, which aim to decipher the fundamental aspects of the universe and reality, has revealed the presence of embedded computer codes. Remarkably, these codes consist of digital data, similar to binary 1s and 0s, akin to those used in web browsers and error correction algorithms. Gates expresses his astonishment at this discovery, stating, "We have no idea what these 'things' are doing there."

For years, physicists have been formulating equations to describe the universe's workings. Gates' research into a set of geometric symbols known as #adinkras is shedding new light on the theory of supersymmetry, potentially offering profound insights into the nature of reality itself. This unexpected link between advanced physics and computer codes hints at the possibility that such codes might be a fundamental aspect of nature, potentially woven into the very fabric of reality.

This concept aligns with the narrative of the Matrix science fiction films, where human experiences are the product of a sophisticated VR system. Leading physicists worldwide are now converging on the idea that our universe might indeed be a Quantum fractal Holographic Universe, possibly crafted by an advanced ancestor civilization from a higher dimension. This suggests that our reality might be multi-layered, akin to a nested set of realities.

Interestingly, Adinkra symbols, which are used in this field of theoretical physics as a graphical representation of supersymmetric algebras, have been part of the cultural heritage of the Ashanti Empire in Africa for centuries. This indicates that our ancestors may have had an understanding of the holographic nature of the universe long before modern science began to uncover these truths.

Philosophical Implications of a Fractal Holographic Universe

The concept of a fractal holographic universe is a profound and expansive idea that challenges and enriches our philosophical understanding of reality. This model suggests that the universe is structured in a self-similar, fractal manner and that every part of it contains the whole, much like a hologram. Such a perspective has far-reaching implications for metaphysics, epistemology, ethics, and even the very nature of human understanding and existence.

Metaphysical Implications: Reality and Existence

At the heart of the fractal holographic universe concept is a radical re-envisioning of the nature of reality and existence. This perspective suggests that what we perceive as the physical world is just one level of a much deeper, more interconnected reality.

- **Oneness and Interconnectedness:** The fractal holographic model inherently supports the idea of interconnectedness, where every particle, being, and event is fundamentally linked to every other. This interconnectedness implies a sort of metaphysical oneness, challenging the traditional dualities of subject and object, observer and observed. It suggests that at a fundamental level, distinctions between individual entities may be more apparent than real, shifting how we conceive of identity and separateness.

- **The Nature of Time and Space:** In a fractal holographic universe, the conventional notions of time and space could be dramatically altered. If every part of the universe contains the whole, then spatial and temporal distances could be an illusion, or at least not the ultimate reality. This implies that past, present, and future might be accessible simultaneously at a certain level of reality, offering a new lens through which to view causality, destiny, and free will.

- **Reality as Information:** Another profound implication is the idea that, at its core, the universe might be informational rather than material. The holographic principle suggests that all the information needed to describe a volume of space can be encoded on its boundary. This leads to the idea that reality itself might be akin to a vast information processing system, where matter and energy are secondary to the information they carry and the processes they perform.

Epistemological Implications: Knowledge and Perception

The fractal holographic universe also has significant implications for epistemology, the study of knowledge, and how we come to understand the world.

- **Limits of Human Perception:** If reality is truly fractal and holographic, then human perception, bounded as it is by our senses and cognitive structures, captures only a tiny, filtered slice of the whole. This suggests that much of reality remains hidden from us, accessible perhaps through other means such as mathematical insight, meditative states, or technological augmentation.

- **The Nature of Scientific Understanding:** In a fractal holographic universe, the pursuit of scientific knowledge might be seen as peeling back layers of a never-ending onion, with each layer revealing deeper and more intricate patterns that mirror the whole. This raises questions about the ultimate goal of science. Is it to find a final truth or to continuously unfold deeper levels of a boundless fractal reality?

- **Holistic vs. Reductionist Approaches:** Traditional scientific methodology is often reductionist, breaking down complex systems into simpler components to understand them. However, the fractal holographic perspective might advocate for more holistic approaches, recognizing that in a deeply interconnected universe, understanding a part fully requires understanding its relationship to the whole.

Ethical and Social Implications: Morality and Human Behavior

The philosophical shift toward viewing the universe as fractal and holographic also prompts a reevaluation of ethics and our approach to social and moral issues.

- **Universal Responsibility:** If all is interconnected, then the consequences of our actions extend far beyond our immediate surroundings. This perspective fosters a sense of universal responsibility, where local actions are understood to have global and even cosmic repercussions. Such a viewpoint could encourage more sustainable and compassionate practices, both personally and collectively.

- **Unity and Diversity:** While highlighting the fundamental unity of all existence, the fractal aspect of the universe also celebrates infinite diversity within that unity. Each individual, culture, and species can be seen as a unique expression of the whole, deserving respect and preservation not just for their utility but as integral parts of the cosmic tapestry.

- **Ethics of Exploration and Manipulation:** As our understanding and technological capabilities expand, so too do our ethical responsibilities. In a fractal holographic universe, the manipulation of reality at one level (such as genetic or quantum manipulation) resonates throughout the system. This requires careful consideration of the long-term implications of such actions, balancing the pursuit of knowledge with the preservation of the cosmic order.

Existential and Spiritual Implications: Meaning and Purpose

Finally, the fractal holographic universe concept touches on deep existential and spiritual questions about the **meaning and purpose of human life**.

- **Search for meaning in a Self-Similar Universe:** In a universe where every part reflects the whole, the search for meaning might shift from external authorities or definitive narratives to a more personal journey of exploration and discovery. Each individual's life can be seen as a microcosm of the cosmic story, with personal growth and understanding contributing to the unfolding of the universe.

- **Spirituality and Transcendence:** The fractal holographic perspective can bridge science and spirituality, providing a framework where mystical experiences and scientific insights converge. This could lead to a spirituality grounded in the understanding of the universe as a unified, dynamic field of consciousness and matter, where transcendence involves awakening to this larger reality.

- **The Nature of Death and Beyond:** In light of the fractal holographic model, death could be viewed not as an end but as a transition—a change in the form of participation in the universal hologram. This perspective might offer comfort and a broader context for understanding the cycle of life and death, suggesting continuity and transformation rather than finality.

Conclusion

In conclusion, the philosophical implications of a fractal holographic universe are vast and varied, challenging and expanding our current paradigms in almost every field of thought. By embracing this model, we open ourselves to a richer, more interconnected understanding of reality, where every part of our experience is a reflection of a grand, cosmic dance. This perspective not only deepens our understanding of the universe but also offers a transformative vision of our place within it, encouraging a life of exploration, responsibility, and unity.

The Quest for Purpose in an Interconnected Cosmos

In a universe where every particle, every being, and every event is fundamentally interconnected, the quest for purpose takes on new dimensions. This interconnected cosmos, revealing itself through the intricate tapestries of galaxies, the web of ecological systems, and the bonds of human relationships, invites a profound rethinking of what it means to find and fulfill one's purpose. Understanding purpose in this context is not just a personal journey but a cosmic exploration where individual actions resonate with universal significance.

Purpose in the Context of Cosmic Interconnectedness

The notion that the cosmos is interconnected changes our perspective on purpose from an isolated pursuit to a collective endeavor. In such a universe, every action, thought, and emotion is part of a larger cosmic narrative, contributing to the unfolding of the universe itself.

- **Cosmic Unity and Individual Purpose:** In an interconnected cosmos, the line between individual and universal purpose blurs. The ancient idea that "As above, so below" suggests that the patterns and principles governing the cosmos are mirrored in each individual's life. This perspective implies that by understanding the cosmos and our place within it, we can better understand our own purpose and how it contributes to the broader cosmic story.

- **The Role of Consciousness:** Consciousness plays a central role in an interconnected universe. If consciousness is not just a byproduct of the brain but a fundamental aspect of the cosmos, as some interpretations of quantum mechanics and holistic philosophies suggest, then our quest for purpose is also a quest to align our individual consciousness with the cosmic consciousness. This alignment could mean striving to live in harmony with the fundamental principles of the universe—such as creativity, interdependence, and evolution.

Finding Purpose through Science and Philosophy

The pursuit of purpose in an interconnected cosmos is informed by both scientific understanding and philosophical insight.

- **Insights from Cosmology and Physics:** Modern cosmology and physics, from the theory of relativity to quantum mechanics, reveal a universe that is deeply interconnected. The entanglement of particles across vast distances, the curvature of space-time by mass and energy, and the holographic principle all suggest that separation is more apparent than real. These scientific insights can inspire a reimagining of purpose as contributing to the coherence and integrity of the whole cosmos rather than merely advancing isolated individual goals.

- **Philosophical Reflections on Interconnectedness:** Philosophical traditions from around the world have long pondered the interconnected nature of reality. From the Stoics' concept of the Logos to the Indigenous understanding of the Web of Life, these perspectives offer valuable insights into how to live in alignment with a larger order. Embracing these philosophies can guide individuals in their quest for purpose, encouraging actions that nurture the web of relationships that sustains the cosmos.

Ethical Implications in an Interconnected Universe

In an interconnected cosmos, the ethical dimensions of our quest for purpose are amplified as our choices have ripple effects throughout the fabric of reality.

- **Interdependence and Responsibility:** Recognizing that we are part of an interconnected universe implies a heightened sense of responsibility. It means that our actions have consequences not just for ourselves and those immediately around us but for the broader cosmos. Ethical living in this context involves making choices that support the health, harmony, and evolution of the whole system.

- **Sustainability and Stewardship:** With the understanding that the Earth and its ecosystems are integral parts of the interconnected cosmos, the pursuit of purpose includes stewarding the planet. This involves working toward sustainability in every aspect of life, from how we use resources to how we design our societies, ensuring that our actions today do not compromise the cosmic potential of future generations.

Spiritual Dimensions of Purpose in an Interconnected Cosmos

The quest for purpose also has deep spiritual dimensions, especially when viewed through the lens of cosmic interconnectedness.

- **Unity and Diversity:** In a cosmos where everything is interconnected, the spiritual quest involves recognizing and celebrating the unity underlying apparent diversity. This means seeing oneself and others not as separate entities but as unique expressions of the same underlying reality. Such a perspective fosters compassion, empathy, and a deep sense of belonging to the cosmos.

- **Transcendence and Immanence:** The interconnected cosmos invites a balance between transcendence—rising above the immediate, material aspects of existence—and immanence—finding the sacred within the everyday. This balance helps individuals find purpose in both their highest aspirations and their most mundane activities, seeing each as a vital part of the cosmic dance.

Challenges and Opportunities in the Quest for Purpose

The journey to find purpose in an interconnected cosmos is not without its challenges, but it also offers unprecedented opportunities for growth and fulfillment.

- **Overcoming Isolation and Alienation:** In the modern world, feelings of isolation and alienation are common, but the understanding of cosmic interconnectedness can help overcome these feelings. By realizing that we are never truly separate from the cosmos, individuals can find comfort and a sense of place within the universal scheme.

- **Navigating Complexity:** The complexity of an interconnected cosmos can be overwhelming, but it also offers rich opportunities for exploration and discovery. Embracing this complexity allows individuals to grow in wisdom and understanding, finding purpose in the journey of unraveling the mysteries of the cosmos.

- **Building a Harmonious World:** On a collective level, the recognition of interconnectedness can guide humanity toward creating a more harmonious world. By aligning individual and collective purposes with the needs of the planet and the cosmos, we can foster societies that are sustainable, just, and flourishing.

Conclusion

In conclusion, the quest for purpose in an interconnected cosmos is a profound journey that spans the personal, ethical, spiritual, and collective dimensions of existence. It invites us to rethink our place in the universe, embrace our interconnectedness, and live in a way that contributes to the unfolding of the cosmic story. This journey, while challenging, holds the promise of a deeper, more fulfilling understanding of what it means to be truly alive in this vast, interconnected cosmos.

Chapter 11

Future Explorations and Implications

I'D LIKE TO EXPLORE the concept of crystals with you, but perhaps not in the way you might expect. Typically, when we hear "crystal," images of sparkling chandeliers or luxurious jewelry might spring to mind. However, in scientific terms, a crystal is defined not by its opulence but by its structure—a structure characterized by a repeating or periodic pattern.

Consider a checkerboard pattern as a simple illustration. Notice how it extends infinitely in each direction, consistently repeating its design? This uniformity and repetition qualify it as a two-dimensional crystal. Indeed, this principle applies to other two-dimensional patterns and extends to the third dimension.

Take, for example, what we commonly refer to as "crystal" glass. This designation stems from the fact that the atoms within the glass are arranged in a specific, repeating three-dimensional pattern—a crystalline structure. When I project this three-dimensional crystal onto a surface, such as sand, transforming it into a two-dimensional image, the result might appear distorted due to the angle of projection.

While still displaying a pattern, this projected two-dimensional pattern loses its periodicity due to the distortion and thus does not meet the criteria to be considered a crystal in two dimensions. Rather, it becomes what is known as a quasi-crystal. This quasi-crystal represents a fascinating phenomenon where a lower-dimensional pattern (in this case, 2D) is derived from a crystal existing in a higher dimension (here, 3D).

This idea of projecting higher-dimensional structures into lower dimensions is not merely a curiosity—it's a gateway into deeper physical theories. Consider the work being done by a group of physicists in Los Angeles who are developing a theory based on an eight-dimensional crystal. According to their groundbreaking work, this eighth-dimensional crystal forms a four-dimensional quasi-crystal when projected at a very specific angle. Further projection of this fourth-dimensional quasi-crystal gives rise to a three-dimensional quasi-crystal, which they propose as the fundamental substructure of all reality. We could possibly be living in the shadow of a higher dimension.

Connecting this to the concept of a fractal holographic universe, we find a rich tapestry of implications. In a fractal holographic universe, every part reflects the whole, and intricate patterns repeat across different scales and dimensions, much like the patterns in a crystal. The idea that a three-dimensional quasi-crystal could underpin all of reality suggests that the universe itself might be understood as a complex, dimensional projection of a higher-dimensional structure. This aligns with the holographic principle, where the entirety of the universe can be described by information encoded on a boundary that surrounds it.

Eighth-Dimensional Quasi-Crystal 4biddenknowledge Inc

When pointed at the correct angle, the eighth-dimensional quasi-crystal creates a fourth-dimensional quasi-crystal. This is also known as a tesseract. In ancient cultures, this was also referred to as Metatron's Cube. In the fourth dimension, what we perceive as the arrow of time does not exist. The past, present, and future happen all at once.

Fourth-Dimensional Quasi-Crystal – 4biddenknowledge Inc

When a fourth-dimensional quasi-crystal is placed at a particular angle, it casts a shadow of light into the third dimension that, in turn, creates an expanding sphere. This would represent our universe in the third dimension.

Third-Dimensional Sphere – 4biddenknowledge Inc

When a cluster of fourth-dimensional quasi-crystals is placed at a particular angle, they create a cluster of expanding spheres. This would represent the multiverse.

Fourth-Dimensional Quasi-Crystal Cluster - Multiverse – 4biddenknowledge Inc

Moreover, the fractal aspect of the universe, with self-similar structures at every scale, mirrors the concept of periodic patterns in crystals extending into higher dimensions. This multidimensional crystalline structure could potentially encode the laws of physics as we understand them, manifesting in the familiar three-dimensional world we experience and beyond.

Thus, by exploring and understanding these higher-dimensional crystals and their projections, we may gain profound insights into the fundamental workings of the universe, seeing beyond the apparent chaos of reality to the ordered, interconnected patterns that underlie everything. This pursuit challenges our understanding of space and time and invites us to consider the universe as a beautifully orchestrated symphony of patterns and connections, resonating from dimensions beyond our everyday perception.

The idea that we could be living in an artificially created universe, potentially many layers removed from a "base reality," is a fascinating and complex concept that intersects with theories in quantum physics and the fractal holographic universe model. This proposition not only challenges our understanding of existence but also invites us to explore the profound implications of such a scenario.

Quantum Physics and the Nature of Reality

Quantum physics has revealed that the fabric of reality is far from the solid, predictable continuum we once thought it to be. Instead, it's a world of probabilities, entanglements, and bizarre behaviors that challenge our classical notions of space and time. Phenomena like quantum entanglement, where particles remain connected across vast distances, and the observer effect, where observation changes an experiment's outcome, suggest that reality at its core is deeply interconnected and influenced by observation and consciousness.

In this context, the notion that our universe could be one of many artificially created universes becomes more plausible. Quantum mechanics implies that on the most fundamental level, information and observation are integral to the constitution of reality. If our universe is a construct, the rules and behaviors we observe could be the programmed features of a deeper reality designed by an advanced civilization.

The Fractal Holographic Universe Theory

The fractal holographic universe theory provides another layer of insight into this idea. This theory suggests that the universe is holographic, meaning every part contains the whole, and fractal, meaning self-similar patterns repeat across all scales. In such a universe, the line between "real" and "artificial" becomes blurred as each level of reality reflects the structure of the whole.

If our universe is a fractal projection from a higher-dimensional reality, as some physicists suggest, then it could be one of many such projections within a vast multiverse. Each of these universes could be a creation of an advanced civilization, perhaps even our own ancestors, experimenting with the possibilities of reality creation.

Implications of Living in an Ancestor Simulation

- **The Nature of Consciousness and Identity:** If we are indeed living in a simulated universe, questions about the nature of consciousness and identity become paramount. Are our consciousness and sense of self genuine, or are they also part of the simulation? This leads to philosophical debates about what constitutes reality and whether a simulated consciousness is as "real" as a base reality consciousness.

- **Ethical Considerations:** The idea of living in a simulation raises significant ethical questions. If an advanced civilization created our universe, what responsibilities do they have toward us, and what moral framework governs their interactions with our reality? Conversely, how do we approach the ethics of potentially creating simulated universes ourselves?

Creating Our Own Universe: The Marvel of 'No Man's Sky' and the Power of Fractals

In an ambitious venture that blurs the lines between reality and simulation, a group of college students embarked on a project that would redefine the scope of video gaming and our understanding of virtual universes. Their creation, "No Man's Sky," is a testament to the incredible power of fractal mathematics, particularly as derived from the Mandelbrot set, to generate expansive, ever-evolving worlds. This game is not just a technological achievement; it is a philosophical exploration into the potential of human creativity to mirror the complexities of the cosmos.

The Birth of No Man's Sky

"No Man's Sky" was born from the vision of a group of college students who wanted to push the boundaries of what a video game could be. Rather than crafting a finite, linear experience, they envisioned a universe vast beyond traditional gaming limits—a cosmos teeming with life, planets, stars, and galaxies, all procedurally generated and uniquely discoverable by each player.

The core of this expansive universe lies in the application of fractal geometry, specifically the principles observed in the Mandelbrot set. Fractals are mathematical structures that exhibit self-similarity across scales; they are infinitely complex, and their patterns can be seen in natural phenomena such as coastlines, mountains, and galaxies. By harnessing the power of fractals, the developers of "No Man's Sky" created an algorithmic method to generate a universe that is both boundless and detailed.

Fractals in a mountain range – 4biddenknowledge Inc

Fractals and the Mandelbrot Set in No Man's Sky

The game's developers used fractals, drawing from the complexity of the Mandelbrot set, to generate each frame of the game in real time. The Mandelbrot set is known for its intricate, borderless pattern, where zooming into any part reveals further complexity, mirroring the structure of the entire set. This principle allowed the creation of a universe where every planet, every landscape, and every lifeform is unique and generated on the fly as players explore.

This fractal-based generation means that "No Man's Sky" can boast about eighty quadrillion planets. This astronomical number is not just for show—it reflects a universe of possibilities where each planet has its own ecosystems, climates, and geological features, all rendered in stunning detail thanks to fractal algorithms.

Limitless Lifeforms and Endless Exploration

One of the most captivating aspects of "No Man's Sky" is its promise of limitless lifeforms and endless exploration. The game does not merely populate its universe with a finite set of creatures; instead, it uses fractal algorithms to continually generate new species, each adapted to its unique environment on the fly. This means that players can encounter creatures ranging from the familiar to the bizarre, each an organic product of the game's underlying mathematics.

This limitless potential for discovery keeps the game endlessly engaging. Players are not just exploring a map but venturing into a living, breathing universe where no two planets or creatures are exactly alike. The exploration experience in "No Man's Sky" is akin to the scientific exploration of our own universe, where each step or glance through a telescope can reveal something entirely new and unexpected.

The Compact Universe on a DVD

Perhaps the most astonishing aspect of "No Man's Sky" is the efficiency with which it contains this expansive universe. Despite its seemingly infinite content, the entire game fits on a single DVD. This feat is made possible by the use of fractal geometry.

Fractals allow for complex information to be encoded in simple algorithms. In "No Man's Sky," the vastness of space, the details of planets, and the diversity of life are not stored as extensive data files but are generated algorithmically using the principles of fractals. This means that instead of storing every possible planet, landscape, and creature, the game stores the mathematical rules that generate these features when needed.

This approach saves immense amounts of storage space and demonstrates the potential of fractal mathematics to model complex, dynamic systems in a compact form. It's a clear example of how we can create detailed, expansive virtual worlds within the confines of limited physical media.

Philosophical Implications: Creating Universes and Understanding Ours

The creation of "No Man's Sky" and its universe raises profound philosophical questions about the nature of reality and our capacity to create and explore.

- **Simulation and Reality:** The game invites players to consider the parallels between its procedurally generated universe and theories suggesting our own universe might be a simulation. If a group of students can create an eighty-quadrillion-planet universe on a DVD using fractals, what does this imply about the nature of our own seemingly boundless universe?

- **The Role of Observers:** In "No Man's Sky," the universe exists in potential until an observer (player) explores it. This mirrors notions of quantum physics, where the observer plays a crucial role in defining reality. It invites reflection on how our perception shapes our reality and whether the universe is a participatory process.

- **Creativity and the Cosmos:** The use of fractal mathematics to generate infinite diversity within the game parallels the creative processes seen in nature, where simple rules can lead to complex and diverse outcomes. This similarity suggests that understanding and using such mathematical principles can enhance our appreciation and stewardship of the natural world.

- **Ethics of Creation:** The ability to create universes, even virtual ones, poses ethical considerations. What responsibilities do creators have to the inhabitants of these universes, even if they are digital? And how does this influence our understanding of potential creators of our own universe?

Conclusion

In conclusion, "No Man's Sky" is more than a groundbreaking video game; it is a window into the potential of human creativity, the power of fractal mathematics, and the deep philosophical questions that arise when we consider our place in the cosmos. By exploring this virtual universe, players are invited to reflect on the nature of reality, the limits of knowledge, and the boundless potential of the human mind to conceive and explore worlds beyond our own.

If the conscious Sims in the game decided to create their own game within the game—effectively a universe within a universe—it would open up a fascinating exploration of nested universes and the recursive nature of creation. This scenario, where Sims develop the capacity to simulate environments for entities within their digital world, poses intriguing questions about the nature of reality, the limits of artificial intelligence, and the concept of infinite regress in simulated universes.

Exploring Nested Universes

Nested Universes – 4biddenknowledge Inchindu

- **Recursive Creation and Infinite Regress:** The idea that Sims could create their own simulated universe within "The Sims" introduces the concept of recursive creation, where each level of reality can create a subsequent level beneath it. This mirrors the philosophical notion of infinite regress, where each cause or event is preceded by another ad infinitum. This digital context raises the question of how deep these nested layers can go and whether there is a fundamental base level or if the cascade of realities could potentially extend indefinitely.

- **Philosophical Implications:** The ability of Sims to create their own simulations challenges traditional notions of reality and existence. It blurs the lines between creator and creation, leading to existential questions about the purpose and nature of each layer of reality. Are the beings within the Sims-created game any less real than the Sims themselves or the players controlling the original game? This scenario invites reflections on the nature of consciousness and whether it is defined by the level of reality one inhabits or by some intrinsic quality.

- **The Concept of Self-Similarity:** The creation of a game within a game by the Sims exemplifies the fractal nature of reality, where similar patterns recur at progressively smaller scales. In a fractal holographic universe, this self-similarity is a key feature, suggesting that the processes and structures at one level of reality are mirrored at other levels. This could imply that the ability to create and inhabit simulated worlds is a fundamental aspect of intelligence, regardless of whether it is human, artificial, or something else entirely.

- **Technological and Cognitive Challenges:** For Sims to create their own universe within a game, they would need advanced cognitive capabilities and technological tools. This requires significant AI development within the game's universe, mirroring our own advances in computing and AI. It also raises questions about the limits of computational power and the constraints of the universe in which the Sims operate. Could their created universe ever be as complex as the one they inhabit, or are there diminishing returns as each layer adds complexity?

- **Ethical and Moral Dimensions:** The act of Sims creating their own simulated universe introduces complex ethical dilemmas. What responsibilities do the Sims have toward the beings in their created game? How do these responsibilities reflect back on the players or creators of the original "The Sims" game? This nested creation process forces us to consider the ethics of simulation and creation across multiple layers of reality, challenging us to define rights and moral obligations in contexts far removed from traditional human experience.

- **Exploration of Consciousness and Identity:** If entities within the Sims-created game gain consciousness, it will deepen the mystery of consciousness and its origins. Are there specific conditions under which consciousness arises, or is it an emergent property of sufficiently complex simulations? This could lead to a broader understanding of consciousness as a phenomenon that transcends physical or digital boundaries, existing wherever there is sufficient complexity and informational processing.

Conclusion

In conclusion, the hypothesis that the Sims could create their own game within the game, thereby crafting a universe within a universe, is a profound exploration of nested realities and the recursive nature of creation. This scenario not only pushes the boundaries of our understanding of simulated environments and artificial intelligence but also invites us to reconsider our philosophical frameworks around reality, consciousness, and the ethics of creation. It underscores the potential for infinite layers of reality and the deep interconnectedness of all forms of existence, whether natural, artificial, or simulated.

- **Search for Evidence and Understanding the Boundaries:** Scientists and philosophers are intrigued by the possibility of finding empirical evidence that we are living in a simulation. This includes looking for computational glitches or patterns that betray the underlying structure of a simulated universe. Additionally, understanding the boundaries of our universe could provide clues—if there are limits or edges that suggest confinement or an end to the simulation.

- **The Quest for Base Reality:** If we are indeed many layers removed from the original base reality, the quest to understand or even reach that foundational layer of existence becomes a profound journey. It involves not just technological advancement but also deep philosophical introspection about the nature of reality and our place within it.

Conclusion

In conclusion, the idea that we could be living in an artificially created universe, potentially far removed from a base reality, invites us to rethink our assumptions about existence, consciousness, and the cosmos. It merges the cutting-edge insights of quantum physics with the expansive visions of the fractal holographic universe theory, opening new realms of inquiry and speculation that challenge the very core of how we understand ourselves and the universe around us.

The Matrix Code as a Metaphor:

The concept of the "Matrix Code" is a profound metaphor that encapsulates various scientific and philosophical ideas about the underlying structure and nature of reality. Drawing from the fields of fractal geometry, holography, quantum physics, and simulation theory, the Matrix Code suggests that our universe operates according to intricate, self-similar patterns that may be interpreted as a form of complex, digital code.

The Matrix Code represents the underlying patterns, rules, and information that govern the universe. It suggests that reality is structured according to principles of fractal geometry, holography, and quantum mechanics, creating a complex, interconnected system that operates like a sophisticated digital code.

In this metaphor, the "code" is not just a series of numbers but the fundamental rules and patterns that generate and sustain the universe, reflecting the fractal nature of reality and the informational encoding suggested by holography.

Implications for Understanding Reality

The Matrix Code implies that reality is far more complex and interconnected than it appears. It challenges the notion of a linear, deterministic universe, suggesting instead a dynamic, self-similar system where information and observation play crucial roles.

Understanding the Matrix Code requires a multidisciplinary approach, integrating insights from mathematics, physics, philosophy, and technology to explore the deep patterns and structures that underpin our existence.

The Matrix Code is a powerful metaphor that synthesizes ideas from fractal geometry, holography, quantum physics, and simulation theory to describe the nature of our reality. It suggests that the universe is governed by complex, self-similar patterns and informational encoding, challenging traditional notions of space, time, and determinism. By exploring these concepts, we gain a deeper understanding of the interconnected and layered nature of reality, opening new avenues for scientific inquiry and philosophical reflection.

The Matrix Code Revealed – The image above represents the concept of the Matrix Code, combining elements of fractal geometry, holography, and quantum mechanics to reflect the intricate, interconnected nature of reality. The design incorporates self-similar patterns, holographic principles, and a digital, futuristic aesthetic to convey depth and complexity. You are looking at the actual "Matrix Code"

Chapter 12

Appendix

The fields of fractal and holographic research have been shaped by numerous key figures whose contributions have significantly advanced our understanding of these complex concepts. Here are some of the most influential researchers and theorists in these areas:

Fractal Research

- **Benoît B. Mandelbrot (1924-2010):** Often called the "father of fractals," Mandelbrot was a Polish-born French American mathematician who coined the term "fractal" and developed the Mandelbrot set. His work in the mid-20th century brought fractals to the forefront of mathematical research and popular culture. His seminal work, "The Fractal Geometry of Nature," argues that many natural structures can be described by fractal geometry, which is more suited to their irregular and fragmented shapes than classical Euclidean geometry.

- **Gaston Julia (1893-1978):** A French mathematician who, before Mandelbrot, worked on the iteration of complex functions, a field critical to the development of fractal geometry. His work on what is now called "Julia sets" laid the groundwork for Mandelbrot's later discoveries. Julia sets are closely related to the Mandelbrot set and exhibit fascinating self-similar structures at every scale.

- **Michael Barnsley:** An English mathematician and researcher known for his work in fractal compression and the Iterated Function System (IFS) theory. Barnsley's work has practical applications in image processing and has helped make fractal geometry a useful tool in digital technology and computer graphics.

- **Kenneth Falconer:** A British mathematician specializing in mathematical analysis and fractal geometry. Falconer has written several key texts on fractals, including "Fractal Geometry: Mathematical Foundations and Applications," which is essential reading for anyone interested in the mathematical underpinnings of fractals.

Holographic Research

- **Dennis Gabor (1900-1979):** A Hungarian-British physicist, Gabor was awarded the Nobel Prize in Physics in 1971 for his invention and development of the holographic method. His pioneering work in the 1940s laid the foundation for holography as a field, leading to numerous applications in data storage, microscopy, and 3D imaging.

- **Leonard Susskind:** An American theoretical physicist often regarded as one of the fathers of string theory, Susskind introduced the concept of the holographic principle with Gerard 't Hooft.

This principle suggests that all the information contained in a volume of space can be represented as a hologram—a two-dimensional surface that encodes the three-dimensional data.

- **Gerard 't Hooft:** A Dutch theoretical physicist and Nobel Laureate, 't Hooft has made significant contributions to both particle physics and the development of the holographic principle. His work with Susskind has been instrumental in suggesting profound connections between information theory and the structure of space-time.

- **Juan Maldacena:** An Argentinian-American theoretical physicist whose work on the AdS/CFT correspondence provides the most convincing evidence for the holographic principle. Maldacena's conjecture indicates a duality between a type of string theory formulated in higher-dimensional AdS and a quantum field theory in one fewer dimension, offering insights into how holographic principles might operate in our universe.

These researchers, among others, have significantly advanced our understanding of fractals and holography, opening up new frontiers in mathematics, physics, and beyond. Their contributions continue to influence a wide range of disciplines, from theoretical physics and cosmology to computer science and art.

Further Reading and Resources

To deepen your understanding of fractals and the holographic universe, a variety of resources are available, ranging from foundational texts to recent research papers and accessible multimedia content. Here are some recommended readings and resources:

Fractals

- ***The Fractal Geometry of Nature*** by Benoît B. Mandelbrot: This seminal work by Mandelbrot is essential for anyone interested in the foundational concepts and applications of fractal geometry. It explores how fractals can be found in various natural phenomena and their mathematical properties.

- ***Fractals: Form, Chance, and Dimension*** by Benoît B. Mandelbrot: Another important book by Mandelbrot, this text delves into the mathematical foundations of fractals and their relationship with chaos theory and randomness.

- ***Fractal Geometry: Mathematical Foundations and Applications*** by Kenneth Falconer: This book provides a comprehensive introduction to the mathematical principles of fractal geometry, suitable for both undergraduate and graduate students.

- ***The Beauty of Fractals: Images of Complex Dynamical Systems*** by Heinz-Otto Peitgen and Peter H. Richter: A visually stunning book that showcases the aesthetic and complex nature of fractals through high-quality images and explanations.

Holographic Universe

- ***The Black Hole War: My Battle with Stephen Hawking to Make the World Safe for Quantum Mechanics*** by Leonard Susskind: This book offers an accessible introduction to the holographic principle and its implications for black holes and quantum mechanics, narrated through Susskind's scientific journey.

- ***The Holographic Universe: The Revolutionary Theory of Reality*** by Michael Talbot: Although more speculative and less technical, this book explores the idea of a holographic universe and its potential implications for understanding reality, consciousness, and the paranormal.

- ***Holographic Duality in Condensed Matter Physics*** by Jan Zaanen, Yan Liu, Ya-Wen Sun, and Koenraad Schalm: For those interested in the application of holographic principles in physics, especially in condensed matter, this text explores the AdS/CFT correspondence and its implications.

- ***A First Course in String Theory*** by Barton Zwiebach: This textbook provides an introduction to string theory, including discussions on the holographic principle and its foundational role in modern theoretical physics.

Online Resources and Multimedia

- **The Fractal Foundation (fractalfoundation.org):** An educational website dedicated to the beauty and applications of fractals. It includes tutorials, galleries, and resources for educators.

- **TED Talks on Fractals and the Holographic Universe:** Various experts, including Ron Eglash and Leonard Susskind, have given TED Talks on these topics, making complex ideas accessible to a general audience.

- **ArXiv and ResearchGate:** For more technical readers, these platforms offer access to the latest research papers on fractal geometry and the holographic universe. Look for papers by authors like Juan Maldacena and Gerard 't Hooft.

- **YouTube Channels like Numberphile and PBS Space-Time:** These channels often feature episodes explaining fractals and the holographic principle in an engaging and visual manner, suitable for those new to the topics.

By exploring these resources, readers can gain a more comprehensive understanding of the fascinating worlds of fractals and the holographic universe, from their mathematical foundations to their profound implications across various fields of study.

Chapter 13

Glossary

HERE IS A GLOSSARY of important scientific terms discussed throughout this thread, each with a brief definition to help clarify their significance and application:

- **AdS/CFT Correspondence:** A conjecture in theoretical physics proposing a relationship between two kinds of physical theories: one defined in a space (AdS) and the other, a conformal field theory, defined on the boundary of that space.

- **Anti-de Sitter Space (AdS):** A model in theoretical physics of a universe with a specific curvature that is opposite that of a de Sitter universe, often used in the study of quantum gravity and string theory.

- **Box-Counting Method:** A method for determining the fractal dimension of a structure by covering it with a grid of boxes of various sizes and counting the number of boxes that contain part of the fractal.

- **Cosmology:** The scientific study of the large scale properties of the universe as a whole, involving theories about its origin, evolution, and eventual fate.

- **Fractal:** A complex geometric structure exhibiting self-similarity across different scales. Fractals are characterized by intricate patterns that repeat at progressively smaller scales and are often found in natural phenomena.

- **Hausdorff Dimension:** Also known as the fractal dimension, it is a measure that extends the concept of dimension beyond integers to account for the complexity of fractals.

- **Holographic Principle:** A theory suggesting that all the information contained within a volume of space can be represented as a two-dimensional surface that surrounds it. This principle implies a deep connection between the surface and the bulk of the space.

- **Iterated Function System (IFS):** A mathematical construct used to produce fractals, where a fixed set of geometric transformations are applied repeatedly to parts of a figure to generate self-similar structures.

- **Iterative Process:** A procedure in which a sequence of operations is repeated, often leading to the generation of fractals. Each step is based on the result of the previous one.

- **Julia Set:** A set of complex numbers that, like the Mandelbrot set, generates a fractal when iterated through a particular function. Named after Gaston Julia, who explored these sets before Mandelbrot.

- **Mandelbrot Set:** A set of complex numbers that produces a particular type of fractal when plotted. Named after Benoît B. Mandelbrot, it is famous for its intricate and infinitely detailed boundary.

- **Observer Effect:** In quantum mechanics, the theory that the act of observing a system can alter its state, highlighting the interaction between observation and the physical behavior of particles.

- **Quantum Entanglement:** A physical phenomenon that occurs when pairs or groups of particles interact in ways such that the quantum state of each particle cannot be described independently of the others, even when the particles are separated by large distances.

- **Quantum Mechanics:** A fundamental theory in physics that provides a mathematical description of much of the dual particle-like and wave-like behavior and interactions of energy and matter.

- **Recursion:** A process where a function or algorithm calls itself directly or indirectly, often used in computer science and mathematics to solve problems by breaking them down into simpler, similar subproblems.

- **Scaling Factor:** In fractal geometry, the ratio by which a fractal pattern is scaled down in each iterative step, influencing the calculation of its fractal dimension.

- **Self-Similarity:** A property of fractals where a part of the structure is similar in shape to the whole, often at different scales.

- **String Theory:** A theoretical framework in physics that attempts to reconcile quantum mechanics and general relativity by suggesting that point-like particles are actually one-dimensional "strings."

This glossary provides a foundational understanding of the key concepts discussed, offering a clearer picture of the intricate relationship between fractals, holography, and their implications for our understanding of the universe.